LITTLE EARTH

LondonFieldworks

LITTLE EARTH
LondonFieldworks

Foreword

GUSTAV METZGER

Just days before this foreword had to be submitted there was a report in newspapers that a man had broken into a vitrine in London's Science Museum and stolen a crystal globe that had belonged to John Dee, the Elizabethan natural philosopher. Associated manuscripts were also removed.

This event sums up so much that is wrong in our world. Human greed, criminality, mindless destruction. And the destructors and polluters are more active day by day. Destruction feeds on destruction, and the killing escalates across the country.

How to behave; surely that is the central issue facing our societies. Inevitably, that is the question facing us, those of us involved in that – loosely termed – art world. And it is of particular relevance for the artists, the organisers, the sponsors and funders of that nexus where art, science and technology overlap.

When time can be measured in dimensions that are incomprehensible, when scientists question existence itself, when inventions and discoveries chase through the internet at breakneck speed; where do we stand?

How are we – in our part of the art world – to respond to this escalating profusion of information, and the pressures that go with it? And with this unstoppable flow of information come developments in technical areas tied into commercial exploitation. A step towards unravelling these difficult questions lies in the concept 'ethics'. Judgements are needed – decisions need to be taken. The view that innovation and its application is necessarily considered a good needs to be challenged. The question needs to be put – good for what; good for whom?

In our society there are wide-ranging discussions on science and technology and the risks and dangers involved; these discussions go on in all media – print, radio, TV. This ferment of discussion leaves the art-science field in an isolated position – a kind of 'ghetto', apparently immune to the swirling debates.

Artists are in a privileged position, and privileges involve responsibility and demand exertions. Where the intellectual and technical underpinning of science and technology is involved, vast areas of expanding knowledge should be brought into the equation. This knowledge will need to be considered in the light of the ethical dimensions, particularly so in the spheres of biotechnology. The patrons and the funding bodies will also be expected to keep pace with these ongoing challenges.

There will be people who would be pleased to see artists engaged within the science-technology field, as a kind of cleanup batallion, mopping up the anger and criticism levelled by society against science and technology, with the Capitalist system as integrator, rather like microbes cleaning up pollution or radioactive waste. Art and artists must not be used as a back-up squad for the failed campaigns of the PUBLIC UNDERSTANDING OF SCIENCE groupings.

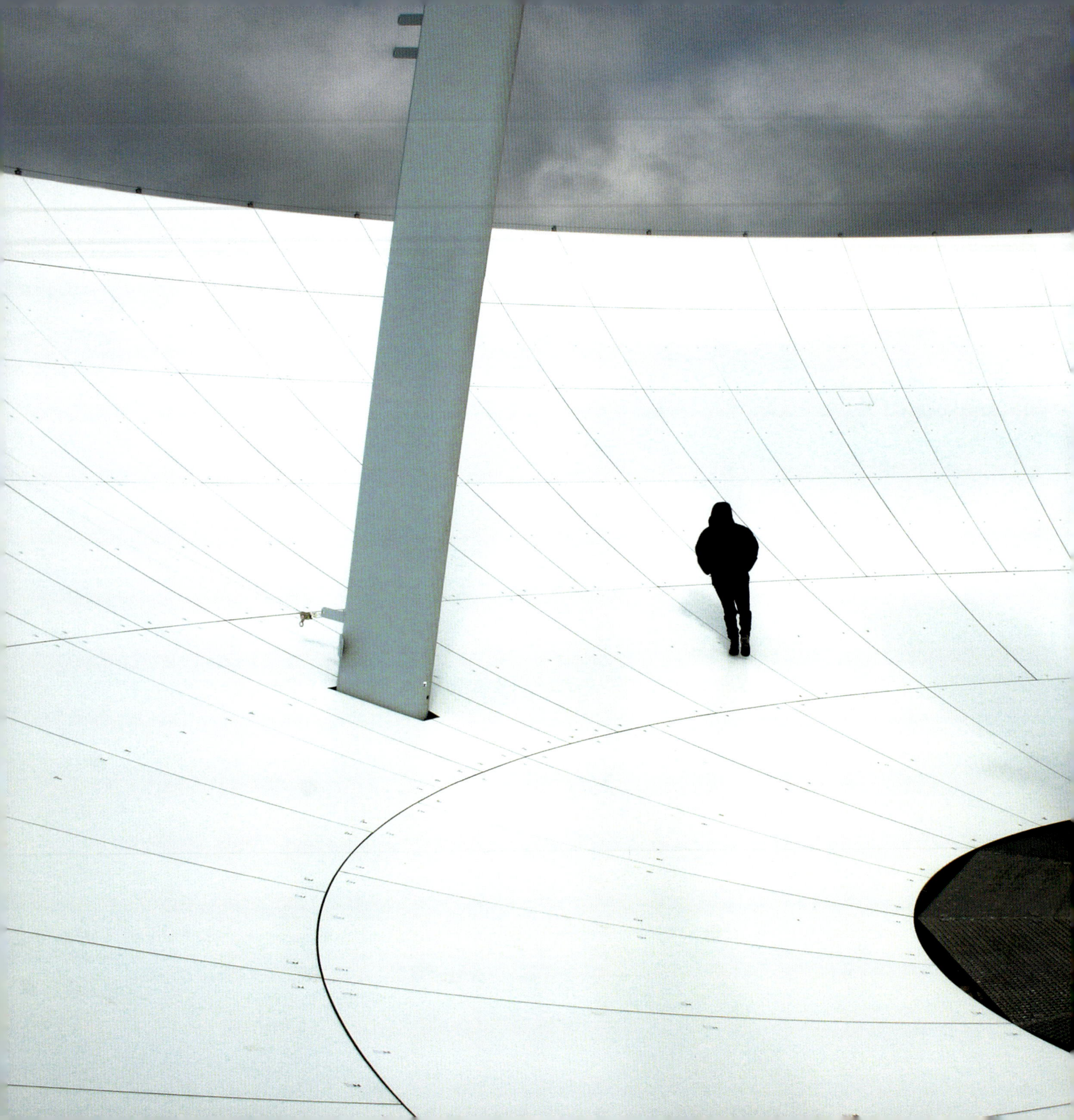

Introduction

BRUCE GILCHRIST & JO JOELSON

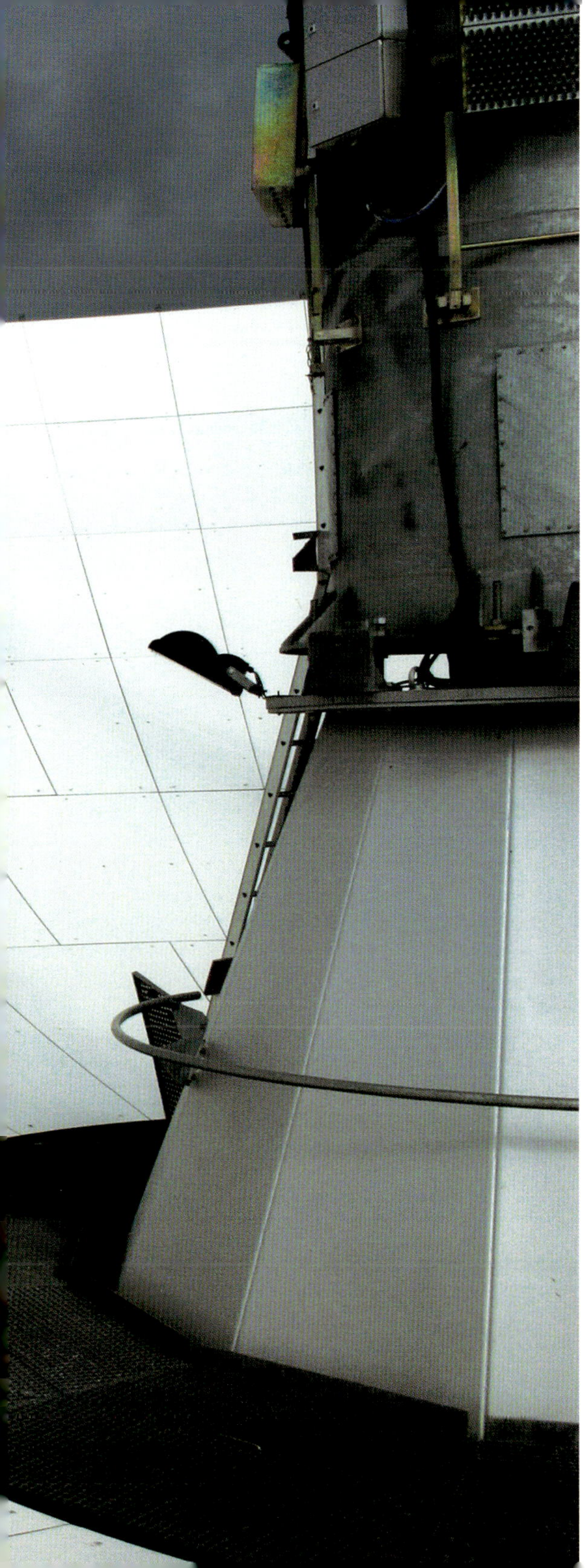

Fixed 42m UHF parabolic antenna
Eiscat Svalbard Radar, Frontiers 8.
Photo: London Fieldworks, 2004

Most of the land surface of the Earth has now been explored. Mankind continues to look towards the mysterious worlds in the sky, reaching further into space.

Over the last century flight has reduced the globe to an accessible sphere. Science has given us machines and instruments that enable us to collect information high above the earth, contributing to what we know about 'space weather'. This knowledge, as well as facilitating off-world exploratory science, influences our perception of the environment in which we live and inevitably determines how we relate to the natural world and what lies beyond our planet.

The *Little Earth* project identifies two physicists, CTR Wilson and Kristian Birkeland, as significant contributors to the development of space weather science. Both researchers were stationed at mountaintop observatories studying natural phenomena at the end of the 19th-century. They relied on naked-eye observation and simple instruments in extreme environments and were perhaps among the last of the natural philosophers. Associated with the mountaintop experiences of Wilson and Birkeland are two celebrated machines: Wilson's cloud chamber and Birkeland's lesser-known terrella. The instruments promoted understanding of phenomena at the micro and macro scale – signalling new frontiers for exploration. These achievements ushered in an era of big science which became increasingly technologised and abstract. The diminutive naked-eye observer standing on the mountaintop was left far behind.

The society in which these scientists lived and worked witnessed the opening-up of vast unimaginable spaces. Revolutionary subatomic-scale experiments and epic explorations of the polar regions provided a correlation between the spaces of late 19th-century experimental physics and uncharted geographical wildernesses.

Using the parallel stories of Wilson and Birkeland as a point of departure we have created a fictional narrative within an artwork to explore the legacies of the two scientists' experiments and instruments. The vehicle for this narrative is itself an experiment in presentation, inspired on many levels by the subject matter: a sculptural object on to which is projected a four-screen synchronised video work incorporating 8-channel sound composition and spoken narrative.

What follows is a collection of writings and documentation charting the development of the *Little Earth* project. Context is provided for the interpretation of London Fieldworks' projects alongside a written and visual document of the ceremony which officially twinned the two mountain observatories. The composer has written an essay describing his response to the *Little Earth* format and his approach to creating the sound work. The *Little Earth* text, written for two voices has been included with a few selected video stills. The historical significance of the two scientists, their instruments and legacies are described authoritatively and in depth by scientists from the fields of meteorology and physics.

This book has been published by London Fieldworks to accompany the exhibition of *Little Earth*.

Our thinking is like lightning in the dark night. It was dark before we started thinking, and it will be dark again when we stop.

Henri Poincar

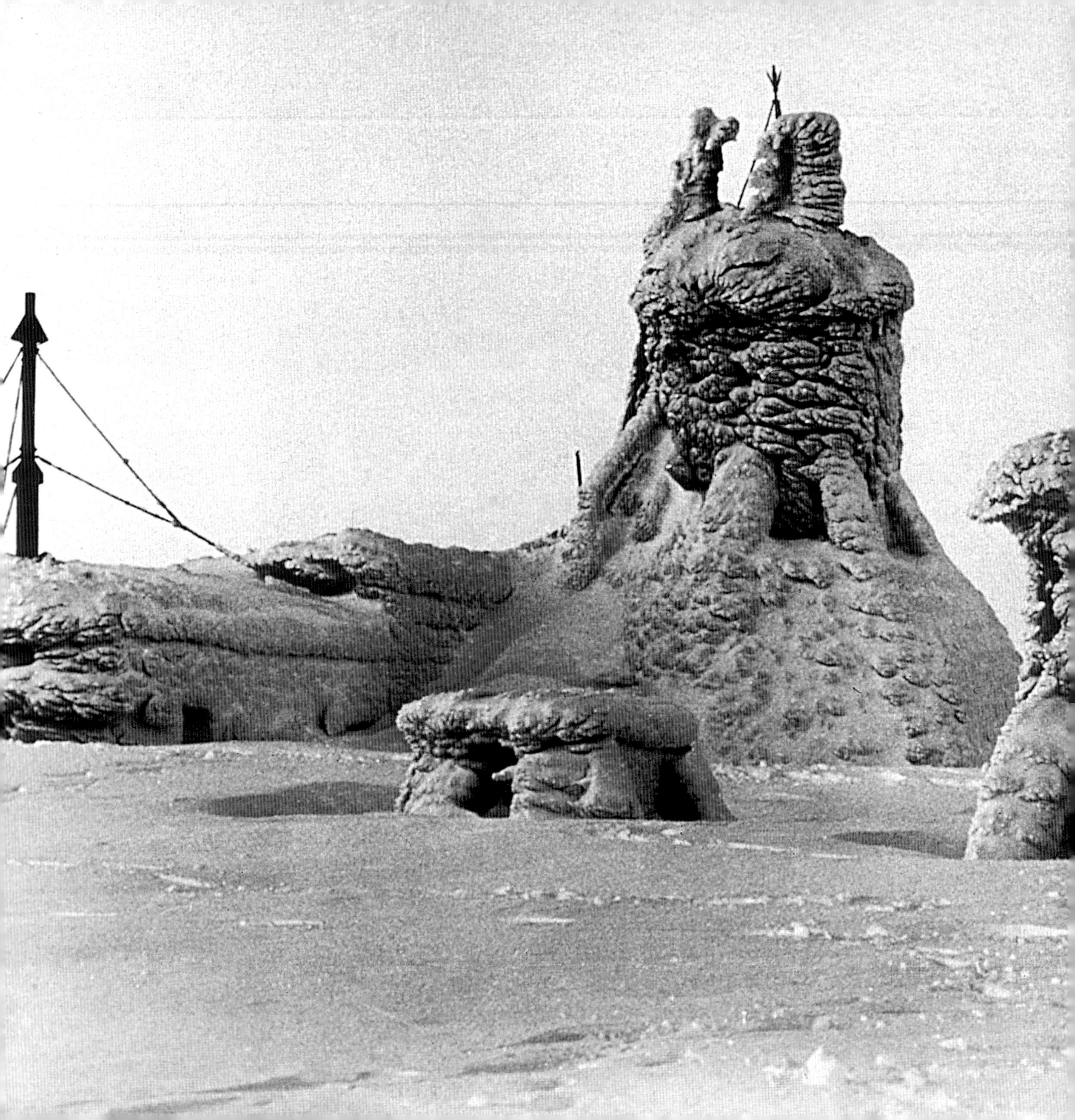

Measuring Beauty In The Upper Ice World

TRACEY WARR

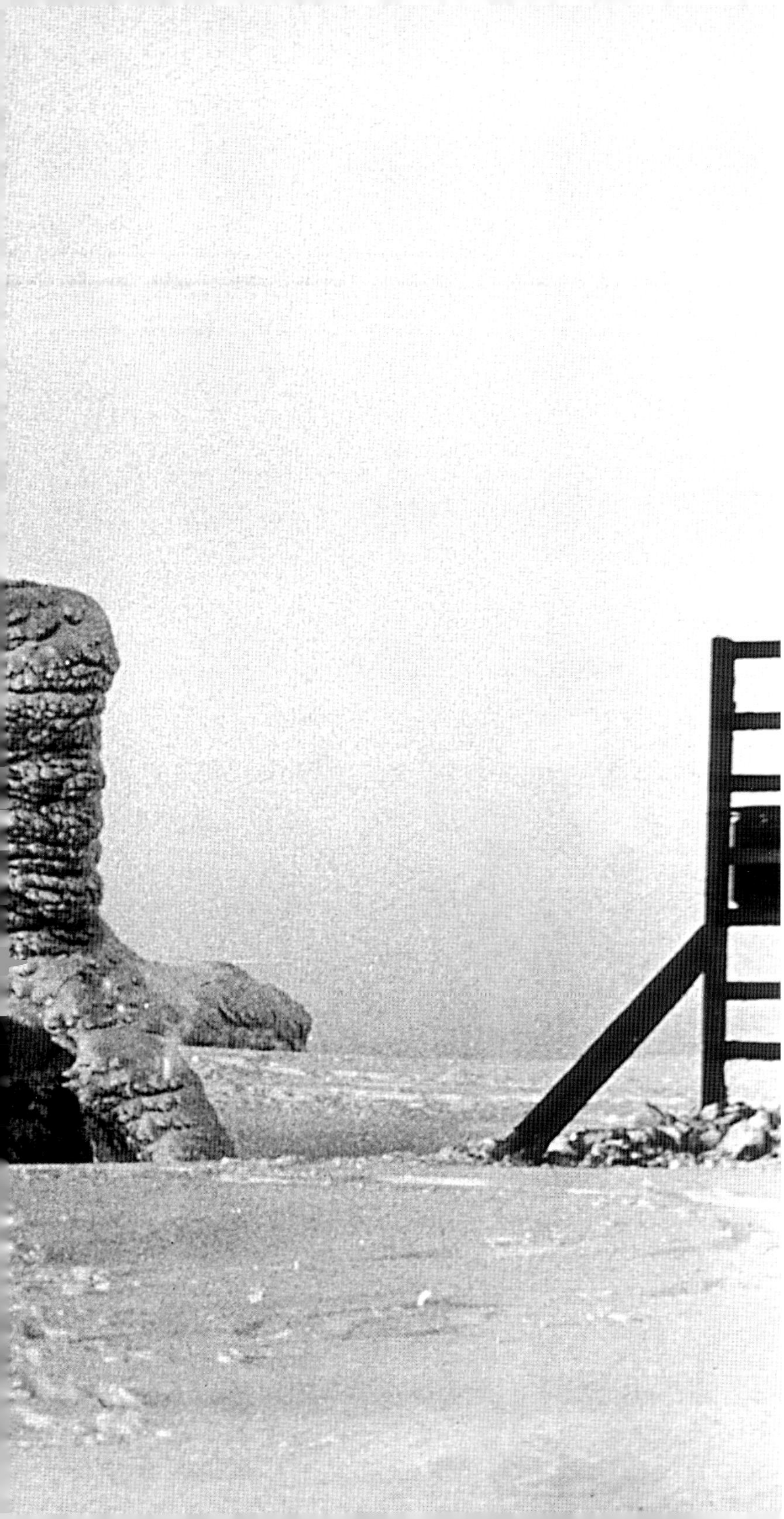

Ben Nevis Observatory Under Snow, c.1890
Courtesy of Edinburgh Met Office

Measurement and Experience

Bruce Gilchrist and Jo Joelson's art project, *Little Earth*, began with two rime-encrusted atmospheric observatories on mountains in Scotland and Northern Norway. The artists researched the lives and work of Victorian scientists, CTR Wilson – who worked at the Ben Nevis Observatory at the turn of the last century investigating electricity in storm clouds, and Kristian Birkeland – who was investigating the aurora borealis at Haldde Observatory. Gilchrist and Joelson engaged with the scientists' experiences by being at the observatories themselves, encountering the weather conditions there and working with the nearby communities in Fort William and Alta. The *Little Earth* project examines the interaction of embodied consciousness with the natural environment. This is a recurring issue in Gilchrist and Joelson's work – in, for example, *Syzygy* (1998/9), *KnoWhere* (2000/1) and *Polaria* (2001/2).[1]

The conditions worked under by Wilson, Birkeland and other meteorologists at the mountain observatories were extreme. In the winter they had to dig their way out of the buildings through piles of snow or heat up instrumentation and hands paralysed by the freezing temperatures. The imminence of an electric storm could be presaged by the experience of having your hair stand on end. There were lightning strikes and battering winds. For long periods the mountaintops and the meteorologists were inside heavily saturated clouds that reduced visibility to a few inches of wet, white, near-tangible fog. Haldde Mountain, above the Arctic Circle, was subject to 24-hour darkness in the winter and 24-hour daylight in the summer. During the summer the meteorologists could experience clear days with endless views of landscape and sea, and balmy temperature inversions when it was hotter at the top of the mountain than at the foot.[2]

The atmospheric phenomena the meteorologists witnessed from the mountains were also extreme. The aurora borealis appeared as spectacular curtains or streamers of green or coloured lights reaching down from the high altitudes of the night sky. They saw glories, haloes, coronae and St Elmo's fire – an electrical discharge that can be seen during thunderstorms around high projecting objects. The Brocken Spectre observed by Wilson on Ben Nevis in 1894 is otherwise known as a 'glory', a phenomenon where the observers see their own magnified shadow thrown on to a cloudbank and encircled by rainbow-like bands.[3]

There were good scientific reasons for undertaking observations on mountaintops. Before the invention of radio it was the only way to investigate the vertical structure of the lower part of the atmosphere and to study the ionosphere – the electrically charged conducting layer in the upper atmosphere – with its electromagnetic forces. Birkeland's research at Haldde linked geomagnetic activity with the aurora. Wilson's observations from Ben Nevis led him to a major contribution to atomic physics and a Nobel Prize.

But there appears to have been some resistance to the mountain observatories from the Victorian scientific establishment. The Ben Nevis Observatory was built and mostly maintained through public subscription and was repeatedly unsuccessful in obtaining any serious financial support from the government or the Royal Meteorological Society. After 21 years of operation it was allowed to fall into disuse, with funding going instead to the Low-Level Observatory in Fort William. Despite the hardships involved in collecting the Ben Nevis weather data sent daily down its heavily armoured telegraph cable, it appears that it was barely used in weather forecasting.

Perhaps part of the resistance to the mountain observatories was an uneasy awareness of their contingency with a romantic engagement with nature. There was intensive activity and interest in the 19th century in both mountaineering and polar exploration, the latter including expeditions to find the Northwest Passage, attempts to reach the North Pole, and to explore Greenland and the Bering Straits. 1882 was the first International Polar Year with 15 polar stations being established. The intrepid 19th-century polar explorers included Ross, Peary, Shackleton, Amundsen and Scott. The glamour of these adventures was only heightened by the fact that the hostile environments took the lives of many mountaineers and polar explorers, including Franklin (1845), De Long (1881), Scott (1912) and their crews. The northern mountains the meteorologists chose to perch on were steeped in associations with the sublime, the mystical, the heroic.

Sublime landscapes are familiar territory in Gilchrist and Joelson's work. Amongst their previous projects *Syzygy* involved fieldwork on the uninhabited Scottish island of Sanda, the *KnoWhere* project took them to Lundy Island in the Bristol Channel, and for *Polaria* they conducted research in remote northeast Greenland. One of their central concerns is an interrogation of the contemporary in relation to Romanticism.

***Polaria* Fieldwork, London Fieldworks**
Northeast Greenland, 2001
Photo: Anthony Oliver

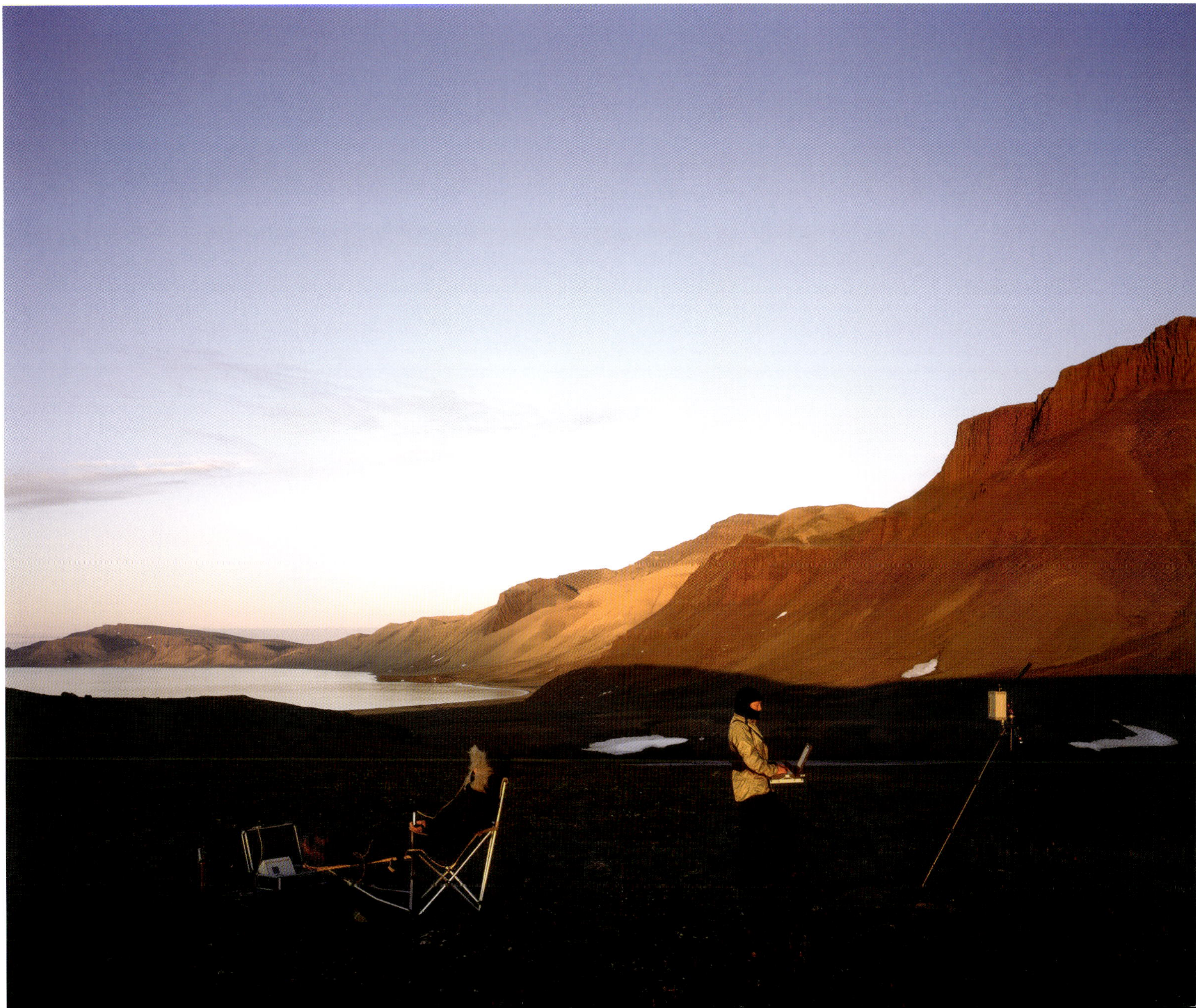

A fascination with awe-inspiring landscapes and natural phenomena – waterfalls, canyons, mountains, storms – developed in the 18th century and was addressed by many writers including Burke, Kant, Goethe and Ruskin. Emerson, Thoreau and Muir also all wrote rapturous essays about mountains.[4] Throughout the 19th century this fascination is imaged in the work of many painters – ranging from the lurid emerald polar seas and threatening icebergs of Frederic Church, to the stormy wrath of God in John Martin's work, the maelstrom of weather and sea in Turner's paintings, and Constable's cloud studies. Many of Caspar David Friedrich's paintings, including The Wreck of the Hope (1823, also known as The Polar Sea) and Wanderer Above a Sea of Fog (1818), emphasise that there was a direct relationship between these artists' work, the activities of gentlemen (and occasionally gentlewomen) explorers and scientists and the dilemma widely felt in European and American 19th-century society in relation to religion and science. Discoveries arising from astronomy and geographic exploration as well as an increasing emphasis on the individual had presented specific challenges to the Church since the 15th and 16th centuries. The impact of the theories of Darwin and other scientists in the 19th century further undermined religious belief. Friedrich's work in particular images a world where divinity is either transferred to nature or is absent altogether and yet nature is redolent with awe nonetheless.[5]

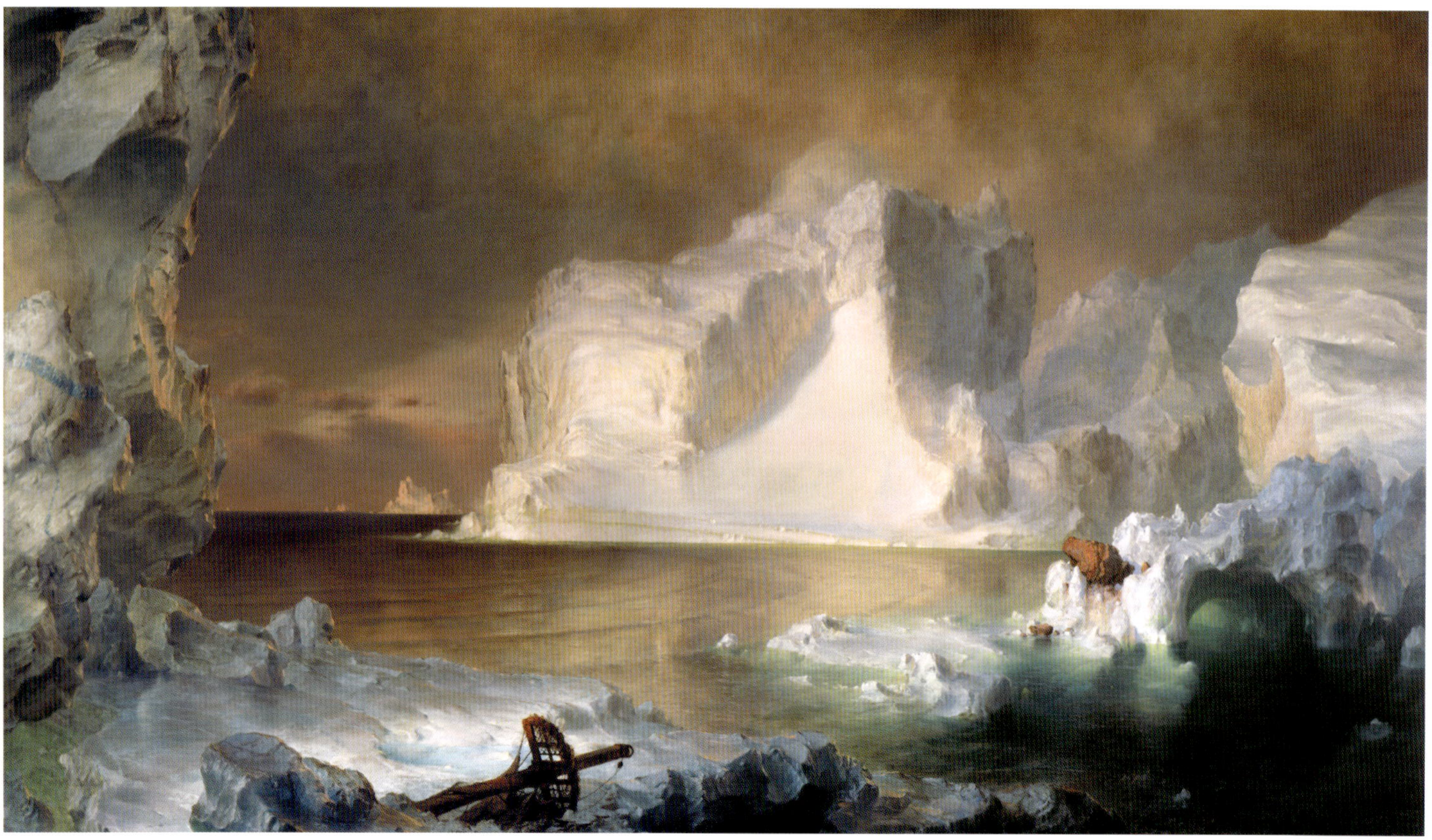

Frederic Edwin Church, The Icebergs, 1861.
Oil on canvas. Dimensions: 64 ½ x 112 ½ in (1 m 63.83 cm x 2 m 85.751 cm)
Dallas Museum of Art, anonymous gift, 1979.28

William Blake had objected to Isaac Newton's measuring of the universe, seeing this as reductive and blind to its wonder. There certainly was a chasm between the handwritten records of hourly observations of temperature, wind speed and direction, and rainfall kept for 21 years at the Ben Nevis Observatory, for instance, and the stunning beauty of the phenomena the meteorologists were witnessing there. There is a gap in articulation between measurement and experience.

Taxonomy, dissection, reduction and disassemblage have been key scientific strategies along with the dualist construct of an objective subject examining an object. Mark Dion's recent work critiques the exercise of power, dominion and closure in a taxonomical methodology. Much of Duchamp's work critiques the reductive idea of knowing through measuring and quantifying. The mathematician and physicist Henri Poincaré had expressed a fundamental doubt as to the possibility of objective scientific knowledge, influencing Duchamp's ideas. Poincaré argued that laws believed to govern matter and its behaviour were created solely by the minds that 'understood' them and that science could not reach the things themselves – only the relations between them.

Traditionally it has been assumed that it is efficacious for the scientist to observe and record data with no reference to the human measurer. Susan Hiller argues that 'objectivity is a fantasy that our culture has heavily invested in'. Thomas Kuhn's theory of paradigms suggests that experiments can produce meanings that are 'contingent on the experimenter's interests'. Subjectivity shapes science, just as it shapes art. Belief, and the limits of belief dictated by the current paradigm, must influence 'knowledge'.

The myth that science works with empirical truths and that art is the unique expression of a born genius is a strong one. Neil deGrasse Tyson, for instance, in an essay entitled 'Science as the Artist's Muse', suggests that

> The most important scientific discoveries...those that came from the minds of undeniably great scientists, would all have been discovered eventually by one or more other scientists... In art, however, Cézanne didn't have to rush-paint his Mont Saint-Victoire out of fear that somebody else was going to create the identical landscape (in Gamwell, 2002: 6).

Both art and science are the result of individuals or teams of individuals working with material found in the collective milieu. Cézanne's innovations were as likely to be, if not replicated, at least approached by other artists, in the same way as a scientist's theories might be.

Despite the overstatement of the notion of construction in some postmodernist theory, the immanence of reality is central to both art and science. Scientific interpretation has been embedded in methodology and has worked towards achieving a stability of knowledge whilst art has always had a capacity to allow fluid interpretations. The understanding in the Humanities is that meaning cannot be definitively deciphered, that interpretation is an unending play with infinitely varied meanings, that art is contingent on its context – the contexts of its framing, the context of its making and of its reception.[6]

Dissection and taxonomy have taken us a long way in the technological and scientific developments of the last two centuries and yet the issues that science is probing now – such as quantum theory or human consciousness – are not yielding at all well to this methodology. There is now a recognition in the scientific community of the value and indeed need for first-person methodologies and an engagement with subjectivity which was hitherto unthinkable, a new understanding of the fluid, dynamic, creative process of interaction going on between subject and object.[7]

Gilchrist and Joelson's work is an ongoing enquiry into the ways that the data of natural phenomena is interpreted and made manifest in both science and art. In *Syzygy* they worked with Imperial College and Cranfield University to develop a smart-materials sculpture that was responsive to physiological and weather data being transmitted to it. In *Polaria* an interactive light installation responded to the physiological state of the visitor. In *Little Earth* they take a different approach – a theatrical engagement with the scientists' experiences – using strategies of re-enactment and staging. In particular they focus on Wilson and Birkeland's invention of idiosyncratic instrumentation, and mirror this in their work with the invention of their own instrument – the *Little Earth* installation.

Instruments of Art and Science

The essays in this publication by Stanley Cowley, Marjory Roy and Alec MacKinnon locate Wilson and Birkeland's work at the turning-point in a shift from naked-eye observation and speculation based in a natural philosophy framework to a science increasingly dependent on instrumentation and technological simulation. These developments show an increasing distance between the scientist and the raw materials or phenomena under examination. Both Wilson and Birkeland invented instruments – Wilson's cloud chamber, which visualised the tracks of ionising particles and Birkeland's terrella machine that demonstrated the aurora and its relationship to solar activity.

Instrumentation and simulation reduce the vast scale of natural phenomena to a human scale where it can be seen, played with (like a doll's house), manipulated and harnessed. The cloud chamber and terrella are in effect models of the atmosphere – the atmosphere in a bottle, a storm in a teacup.

The four-screen video installation of *Little Earth* is also an instrument for understanding and manipulation. The work is a fictional re-creation of the activities of the two scientists at the observatories. The artists describe this as an audiovisual poem. Incorporating a script by James Flint and sound score by Dugal McKinnon, the video work is projected on to the four sides of a cube suspended from an architectural rigging. The projection structure was designed by architect Ed Holloway. He was inspired by radar dishes and antenna arrays the artists had photographed at scientific installations on the island of Svalbard.

Little Earth, however, is not simply the art installation shown at Wapping, Fort William and elsewhere. Gilchrist and Joelson's work characteristically employs a process of interrogating their subject through several modes of artistic 'output'. In *Little Earth* this process has included a series of residencies at the Headlands Center for The Arts, San Francisco, Allenheads Contemporary Arts in Northumberland and at Dartington Gallery in Devon; Joelson's Art/Science Fellowship working with Prof Stanley Cowley and the Radio and Space Plasma Physics Group at the University of Leicester; an official twinning ceremony for the Haldde and Ben Nevis Observatories held at Fort William in October 2004 and this publication in which the artists have collaborated with a number of writers.

During the Headlands Center for the Arts residency the artists made a series of visualisation experiments to represent the methods and ideas of Wilson and Birkeland. They were inspired by the vagaries of the San Francisco microclimate, from the fogs rolling through the Headlands and the Bay Area to temperature inversions at the top of nearby Mount Tamalpais. They were given access to contemporary Space Weather Science for the first time at UC Berkeley Space Sciences Lab and were introduced to the phenomenon of Sprites.[8]

At Allenheads Contemporary Arts, Gilchrist and Joelson conducted a series of interviews with particle and quantum physicists about how contemporary scientific knowledge is disseminated to the lay public. They became interested in the possibility of fictions being created through the interpretations of the lay imagination.

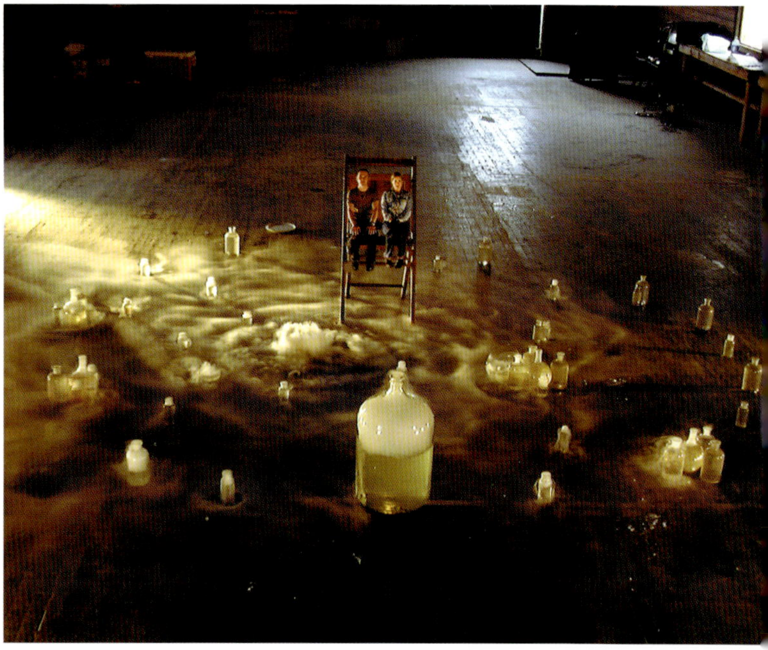

Studio 961, Headlands Center for The Arts, San Francisco 2002
Photo: London Fieldworks

At Dartington they experimented with ferro-liquids and magnetic materials in a lab-like construction of glass vessels and clamps and invented 'proto-instruments for the subconscious'. At each residency location they were at pains to engage audiences in their developing ideas.

Their work is collaborative in a wide-ranging and interdisciplinary way – involving, for example, members of the mountain communities of Haldde and Ben Nevis, a composer, scriptwriter, architect and writers for *Little Earth* and a team including stunt-kite flyers, computer programmers, materials scientists, writers and musicians for Syzygy. Their work is also participatory and it is unusual for artists' practice to span both the gallery world of the art cognoscenti and a range of localised communities. The twinning of the observatories was initiated by Gilchrist and Joelson. They designed the official twinning document with its swathe of aurora across the sky between the mountaintop observatories.

They brought together a range of people from the two remote communities to make it happen, further described in Jeni Walwin's essay. Gilchrist and Joelson's working process has enabled the reconnection of aspects of history and identity for these two remote communities and been a catalyst for a new dialogue: the Director of the John Muir Trust told the community of Alta, for instance, 'Our mountain is now your mountain'.

A lot of current 'sci-art' – where artists collaborate with scientists and work with the concerns and materials of science – is still mired in the idea that art can be an illustrative, accessible, user-friendly mediation for science or that science offers art an alluring range of kit and language to be appropriated. There is still a mutual lack of understanding of functions and methodologies in many science-art projects and above all a misunderstanding of the role of art. Gilchrist and Joelson's work bypasses the binary fallacy of sci-art. Rather than attending to agendas proposed by the artificial construct that is the 'science-art' domain, the artists instead see it as 'part of their creative freedom to measure using their own scale and methods, inspired perhaps by the rigours and questioning of science but unsatiated by its self-imposed limitations'. Cultural authority currently resides with science and not with contemporary art. A puritanical or sceptical streak in traditional science tends to evince a fear of the frivolous and pointless, a fear of the credulous and superstitious. Gilchrist and Joelson's work approaches a complex knot of belief, desire, creativity and knowledge. It reminds us of the play, the intuition, whims and idiosyncrasies in both art and science.

By walking in the scientists' shoes, and through a process of theatrical fictional re-enactment, they have been able to image the intuitive leaps of the scientists and the ways in which they were inspired by the beauty of natural phenomena, and then to relate that to the contemporary world. Whilst they engage aspects of the mimetic and the simulated their work also emphasises the direct engagement of being and becoming. They explore both direct and mediated experience in their work. They enact the role of the generative human imagination in making and unmaking the world and examine how ideas leak out into the manifest world.

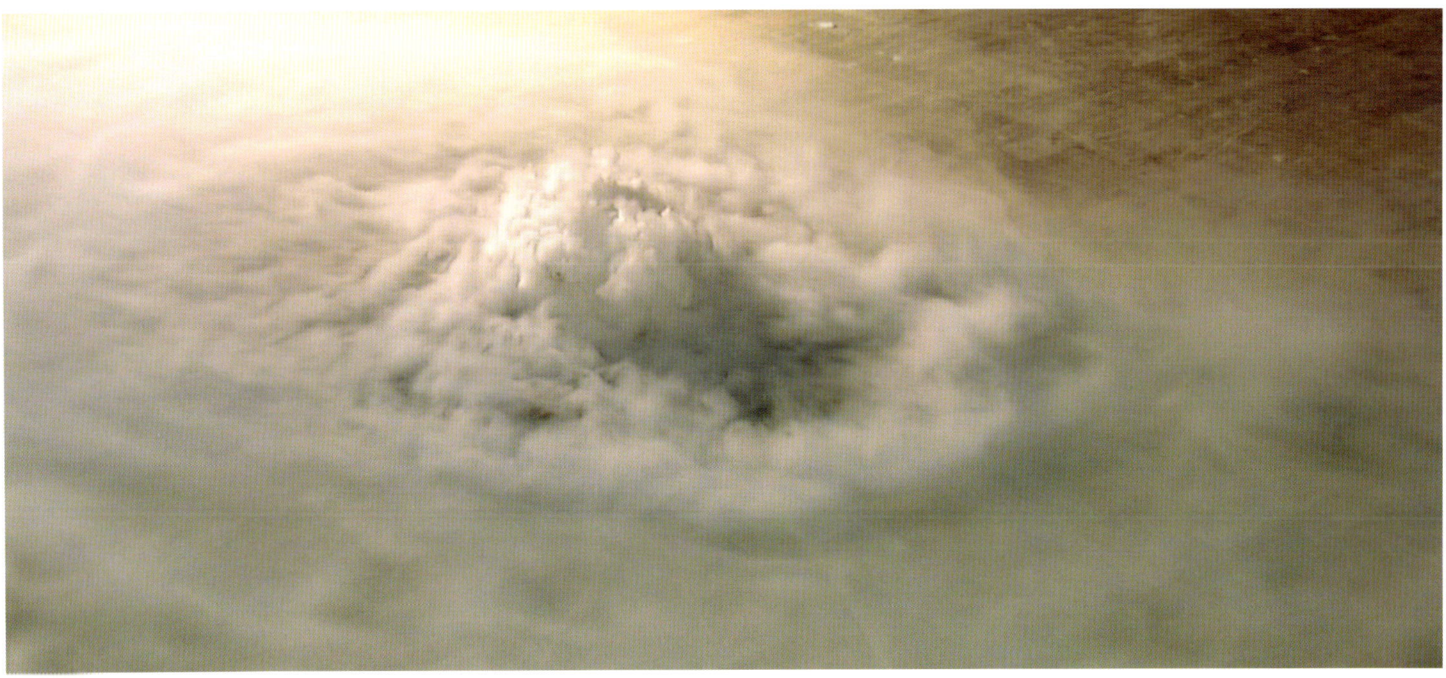

Notes

1. *Syzygy* interrogated the relationship between consciousness and weather; *KnoWhere* was a collaborative work with blind and partially sighted artists exploring sensory modalities in interaction with the natural environment and *Polaria* was concerned with embodied consciousness and natural light phenomena. See Gilchrist, Bruce & Joelson, Jo, eds. (2001). *Syzygy/Polaria.* London: Black Dog Publishing, Gilchrist (2001) CD-ROM, and Warr (2001) article both on *KnoWhere* published in Performance Research, and the London Fieldworks website: http://www.londonfieldworks.com.

2. For a graphic description of the physiological effects of cold and altitude see Ashcroft, Frances (2001). Life at the Extremes. London: Flamingo; and for a description of the meteorologists' life on Ben Nevis see Roy, Marjory (2004) The Weathermen of Ben Nevis 1883-1904. Fort William: Royal Meteorological Society.

3. Brocken is the highest peak in the Harz Mountains in northern Germany. It is reputed to be the scene of witches' Walpurgis-night revels. The relative position of sun, observer, mist and the size of the raindrops and the reflections and refractions of the sun's rays within the cloud droplets produce the Brocken Spectre phenomenon.

4. 'The upper ice-world' is a quotation from Ruskin's Peaks, Passes and Glaciers (1859). See also Macfarlane, Robert (2003). Mountains of the Mind: A History of a Fascination. London: Granta.

5. See Gamwell, Lynn (2002). Exploring the Invisible: Art, Science and the Spiritual. Woodstock: Princetown University Press for a discussion of Romantic art in relation to science.

6. See Ede, Sian (2000). Strange and Charmed: Science and the Contemporary Visual Arts. London: Calouste Gulbenkian and Wilson, Stephen (2002). Information Arts: Interfaces of art, science and technology. London/Cambridge, Mass.: MIT Press.

7. See the Journal of Consciousness Studies and Towards a Science of Consciousness conferences held at the University of Arizona, http://www.consciousness.arizona.edu and abstract of a paper by Gilchrist & Warr (2000) on art and first-person methodologies.

8. Part of the artists' enquiry included discussions with a Stanford researcher who is an authority on the recently imaged sprite phenomena – upper atmospheric, electrical discharges occurring above stormclouds. She created a linkage with CTR Wilson, relating how he predicted the existence of sprites in the 1920s. They were photographed for the first time in 1989. A sprite event has been linked with the 2003 Columbia shuttle disaster: a photograph taken by a San Francisco astronomer appeared to show a purplish bolt of lightning striking the shuttle during re-entry.

References

Ashcroft, Frances (2001). Life at the Extremes: The Science of Survival. London: Flamingo.

Ede, Sian (2000). Strange and Charmed: Science and the Contemporary Visual Arts. London: Calouste Gulbenkian.

Gamwell, Lynn (2002). Exploring the Invisible: Art, Science and the Spiritual. Woodstock: Princetown University Press.

Gilchrist, Bruce & Warr, Tracey (2000). Art as a First-person Methodology in Consciousness Research. In Sutherland, Keith. ed. (2000). Consciousness Research Abstracts: Toward a Science of Consciousness 2000. Thorveton/Tucson: Journal of Consciousness Studies/University of Arizona, p.162.

Gilchrist, Bruce & Joelson, Jo, eds. (2001). *Syzygy/Polaria.* London: Black Dog Publishing.
Gilchrist, Bruce (2001). *KnoWhere.* CD ROM published in Performance Research, 6 (3).
London Fieldworks website: http://www.londonfieldworks.com

Macfarlane, Robert (2003). Mountains of the Mind: A History of a Fascination. London: Granta.

Roy, Marjory (2004). The Weathermen of Ben Nevis 1883-1904. Fort William: Royal Meteorological Society.

Warr, Tracey (2001). Circuitry. In Performance Research, 6 (3), pp.8-12.

Wilson, Stephen (2002). Information Arts: Interfaces of art, science and technology. London/Cambridge, Mass.: MIT Press.

VENNSKAPSERKLÆRING UTSTEDT AV LONDON FIELDWORKS

CERTIFICATE O

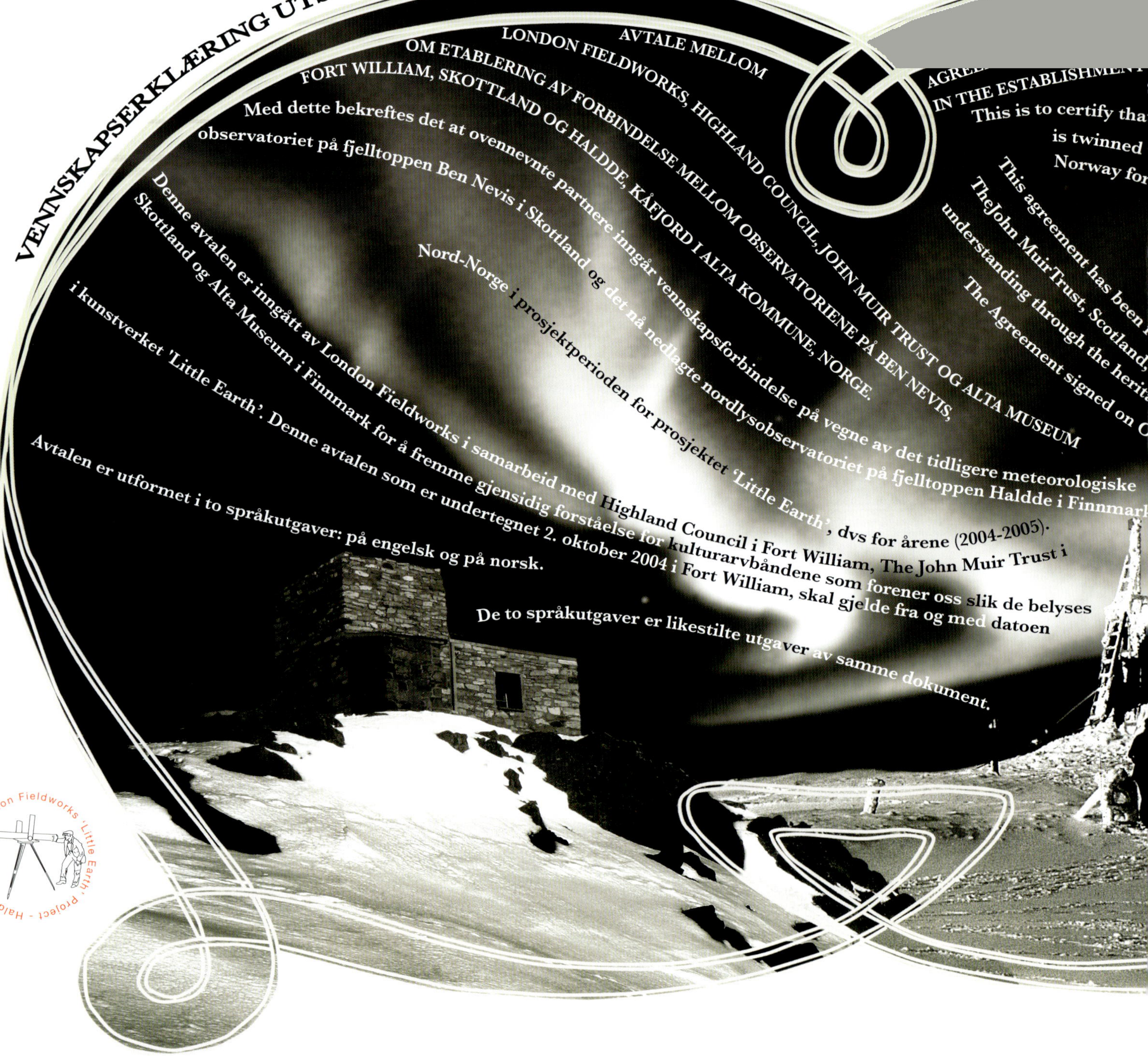

AVTALE MELLOM

LONDON FIELDWORKS, HIGHLAND COUNCIL, JOHN MUIR TRUST OG ALTA MUSEUM

OM ETABLERING AV FORBINDELSE MELLOM OBSERVATORIENE PÅ BEN NEVIS,

FORT WILLIAM, SKOTTLAND OG HALDDE, KÅFJORD I ALTA KOMMUNE, NORGE.

Med dette bekreftes det at ovennevnte partnere inngår vennskapsforbindelse på vegne av det tidligere meteorologiske observatoriet på fjelltoppen Ben Nevis i Skottland og det nå nedlagte nordlysobservatoriet på fjelltoppen Haldde i Finnmark, Nord-Norge i prosjektperioden for prosjektet 'Little Earth', dvs for årene (2004-2005).

Denne avtalen er inngått av London Fieldworks i samarbeid med Highland Council i Fort William, The John Muir Trust i Skottland og Alta Museum i Finnmark for å fremme gjensidig forståelse for kulturarvbåndene som forener oss slik de belyses i kunstverket 'Little Earth'. Denne avtalen som er undertegnet 2. oktober 2004 i Fort William, skal gjelde fra og med datoen

Avtalen er utformet i to språkutgaver: på engelsk og på norsk.

De to språkutgaver er likestilte utgaver av samme dokument.

AGRE

IN THE ESTABLISHMEN

This is to certify tha

is twinned

Norway for

This agreement has been

TheJohn MuirTrust, Scotland,

understanding through the herita

The Agreement signed on O

London Fieldworks 'Little Earth' Project - Haldde Mountain

....
For Alta Museum

....
For Highland Council

....
For John Muir Trus

Twinning Little Earth - The Unfolding of a Ceremony

JENI WALWIN

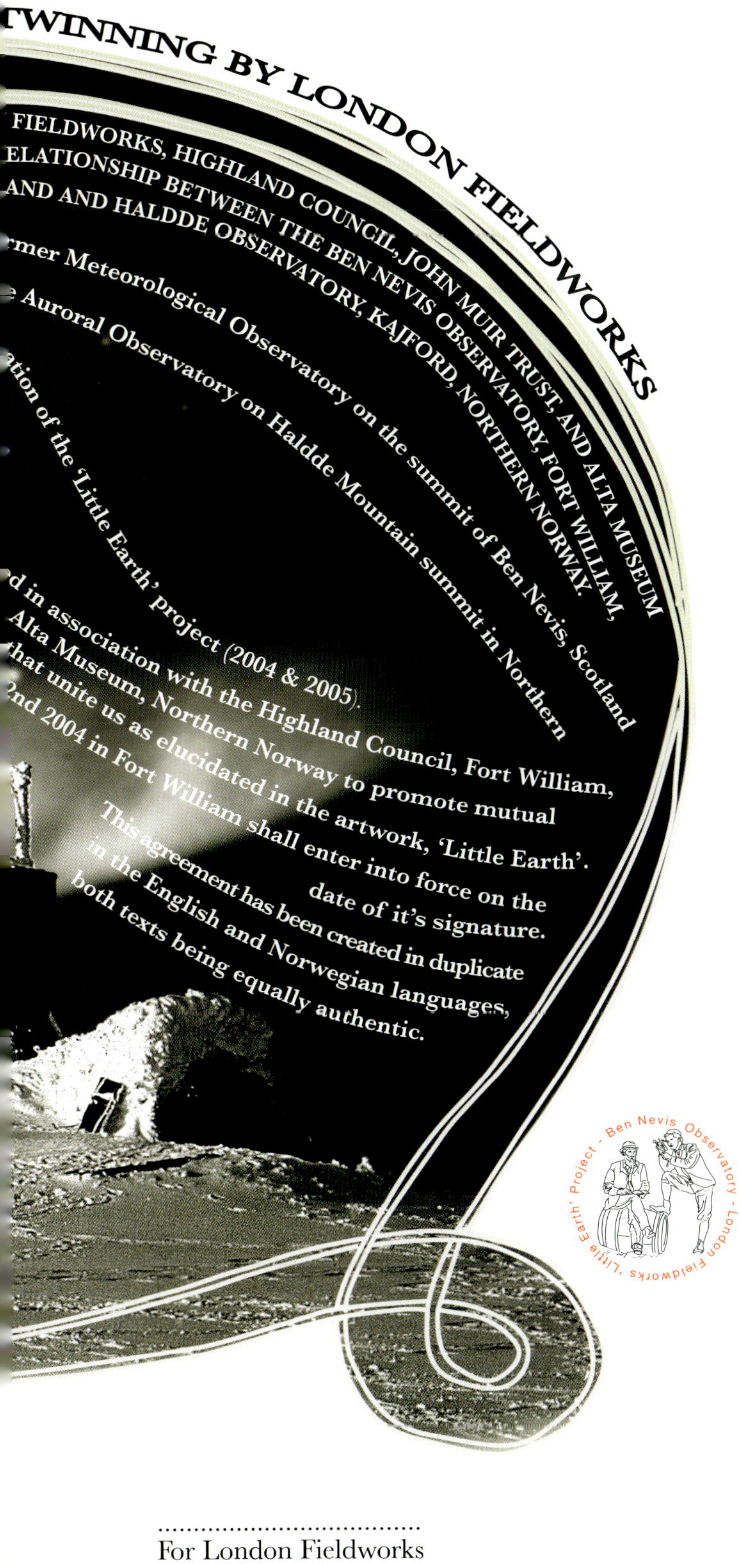

Twinning Agreement, London Fieldworks, 2004

The mountains of Ben Nevis and Haldde encircle the *Little Earth* project at every point in its making. Arriving in Fort William for the twinning event, the shadow of Britain's highest mountain seems to spread its purple cloak over the windswept streets and waterfront buildings. In every corner of this small town and its surrounding villages the presence of the mountain can be felt. It's as if the geological force of Ben Nevis has squeezed the buildings, and all human life associated with them, into the small area of flat land perched between its lower slopes and the shores of Loch Linnhe.

As a subject for artistic enquiry, mountains are loaded. It is hard to imagine that up until the eighteenth century, mountains were considered vulgar and of little significance within western culture. Only with the arrival of travel, communications and scientific development did mountains begin to exert a strange hold on the human imagination. They now infiltrate our everyday experience as exotic travel destinations and an advertiser's dream, symbolising the yearning for the unassailable. Within art history mountains are synonymous with Romanticism and an idealised view of nature. To challenge both the prosaic and conservative rendering of these vast geological outcrops, contemporary artists have proposed ways in which the problematic potential of the subject can be creatively sidestepped, by tracing the evidence of a particular moment and translating it into a new source of experience.

*

I was invited to attend a ceremony at the West Highland Museum in Fort William to mark the twinning of two mountain observatories – the Ben Nevis Weather Observatory and the Northern Lights Observatory on Haldde Mountain in northern Norway. Held on 2 October 2004, exactly a hundred years and a day after the closure of the Ben Nevis Observatory, this event was the first of a two-part work entitled *Little Earth* conceived by artists Bruce Gilchrist and Jo Joelson (London Fieldworks).

The Museum closed its doors to the public for the afternoon of the twinning ceremony and inside the maze of tiny galleries an extraordinarily diverse collection of interests were represented by the guests pressed against the polished mahogany display cases. In the education room the story of the Ben Nevis Observatory and the scientists who worked there was told in slides, photographs, notebooks and letters. Dr Andrew Conway, from the University of Glasgow, demonstrated an original cloud chamber[1] instrument. Local musicians Broslum greeted guests in the Jacobite room with Gaelic folk tunes that seemed to weave their way amongst spinning-wheels, intricately knitted shawls that can be pulled through the eye of a wedding ring, and a finely hewn log boat hauled out of Loch Treig. Tots of Ben Nevis whisky and oatcake canapes were served from an oak birching table that had been in use locally until 1948. Artists, politicians, educationalists, meteorologists, scientists, environmentalists, musicians, journalists, arts funders, local authority officers, mountaineers, photographers, museum curators, filmmakers, actors, local sponsors of the project, their families and assorted children gathered to listen to the speeches and witness the twinning of the two mountain observatories.

Councillor Olwyn Macdonald, Area Convenor of the Highland Council, opened the ceremony, which highlighted the parallels between these two historically significant observatories. The artists introduced the project outlining the inspiration behind it:

> "Both observatories are steeped in history and have been visited by many, including in 1902 the Scott and Amundsen teams in preparation for their respective expeditions to the South Pole. Whilst this demonstrates the significance of the observatories in the late nineteenth century, it is two earlier, less-celebrated figures who initially led us to make the link between the two sites. CTR Wilson and Kristian Birkeland were fuelled by the scientific desire to pry from the earth its long held secrets. For us they have provided a pivotal connection between the past and the present." Jo Joelson

The observatories' influential role in the study of weather and the stories of the scientists who had worked there were outlined by Marjory Roy, the Hon Secretary of the Scottish centre of the Royal Meteorological Society and Hans Christian Søborg, director of the Alta Museum[2] in Norway. The Ben Nevis Observatory functioned between 1883 and 1904 and was devoted to the study of meteorology. Stationed there as a young scientist, and observing his shadow falling on to clouds beneath him, provided the early inspiration for CTR Wilson's cloud chamber. This instrument became a crucial tool in his later work on sub-atomic particles for which he was awarded the Nobel Prize for Physics. In Norway Kristian Birkeland studied the northern lights from the observatory he had built on Haldde[3] Mountain in 1899. His observations led to the construction of the terrella machine which could mimic the creation of aurora phenomena.

The experience of working in each observatory would have been very different. On Ben Nevis the living conditions were rudimentary, whereas the Norwegians enjoyed facilities and services to which the rest of the population had no access. Norway's first telegraph system was made available to the Haldde scientists who lived in the observatory with their families, often bringing pianos and other luxuries up the mountain, and creating a real sense of community in the sky. Norway did much to publicise the scientists' experiments, linking the project to the Norwegian flag at a critical period in the bid for the country's independent status. The Ben Nevis Observatory was closed in 1904 for lack of funding. The Northern Lights Observatory was burned down during the Second World War but was restored in 1983.

Although it is not known if they ever met or even whether they were aware of each other's experiments, more than one hundred years after their initial researches in their remote observatories Birkeland and Wilson have been brought together through the twinning ceremony. *Little Earth* unpicks a particular approach shared by the two scientists, characterised by the proximity they had to their subject, experiencing evidence at first hand and recording lived moments using hand-made instruments.

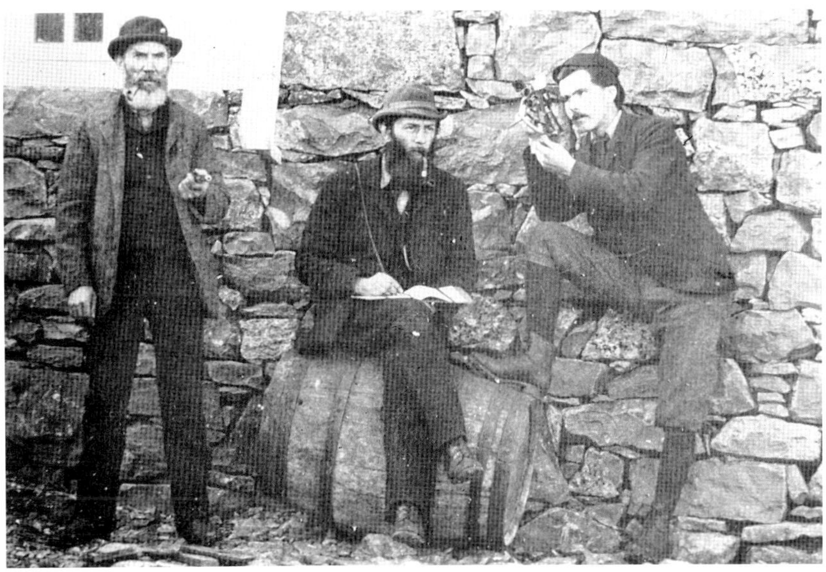

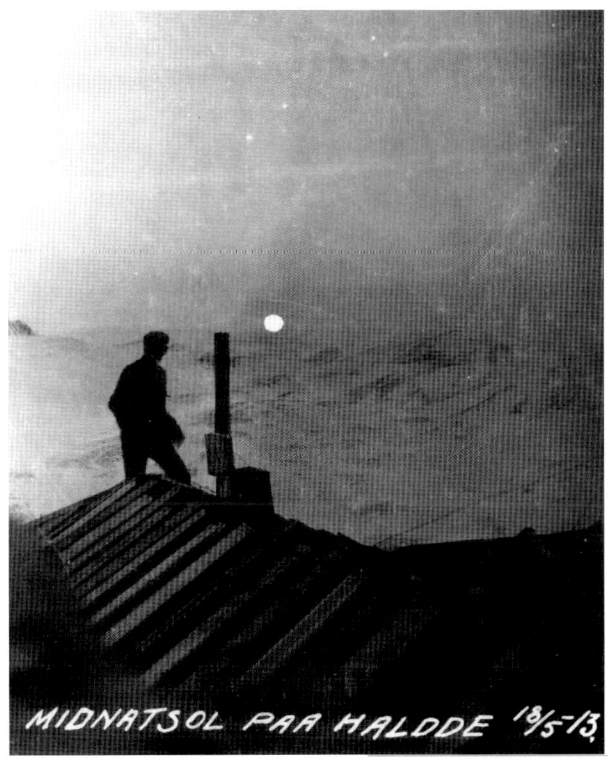

Above: **Observers at Ben Nevis Observatory**, c.1885
Courtesy of Edinburgh Met Office

Right: **Midnight sun at Haldde Observatory**, c.1900
Courtesy of Alta Museum, Norway

At the locations of both observatories, music plays a central role in the culture. Bringing together musicians dressed in costumes of the Sami and Gaelic traditions immediately made the idea of twinning feel more real. Marit Hætta Øverli performed two *joiks* – a *joik* is one of the traditional Sami song-styles. In her introduction she was careful to emphasise that these songs are not stories about an experience but are the experience itself. The deep sonorous moments and joyful calls of her second *joik* transported us to Haldde Mountain, celebrating the particular quality of the northern lights and filling the museum with sparkling echo and low reverberation. By comparison the bagpipes, led by Ali Macdonald, needed the dark hills and open streets to demonstrate their unstoppable ability to override the loudest competition from outbursts of Scottish weather.

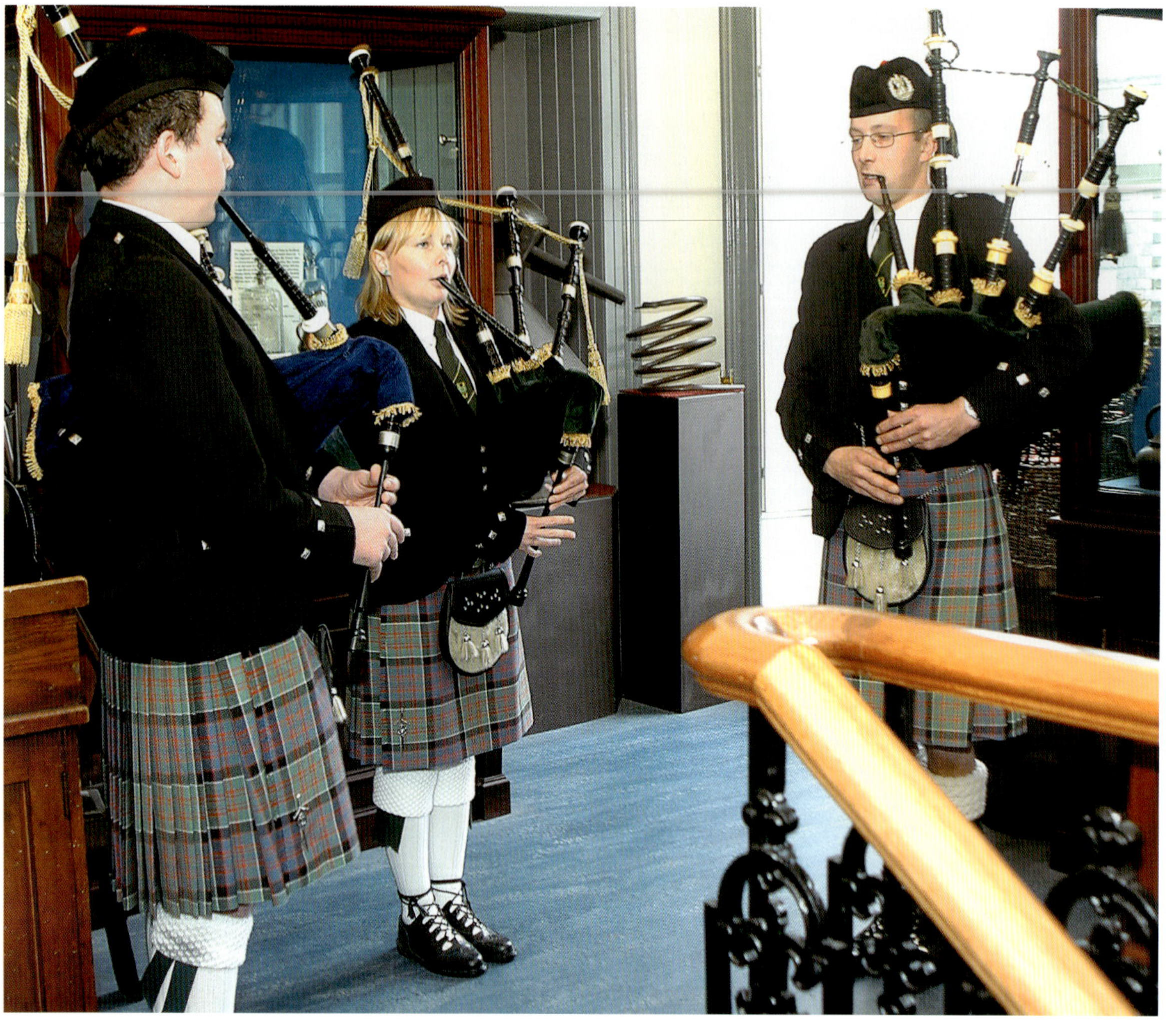

Members of the Lochaber Pipe Band, at the *Little Earth* twinning ceremony, West Highland Museum, Fort William, 2 October 2004.
Photo: Alex Gillespie

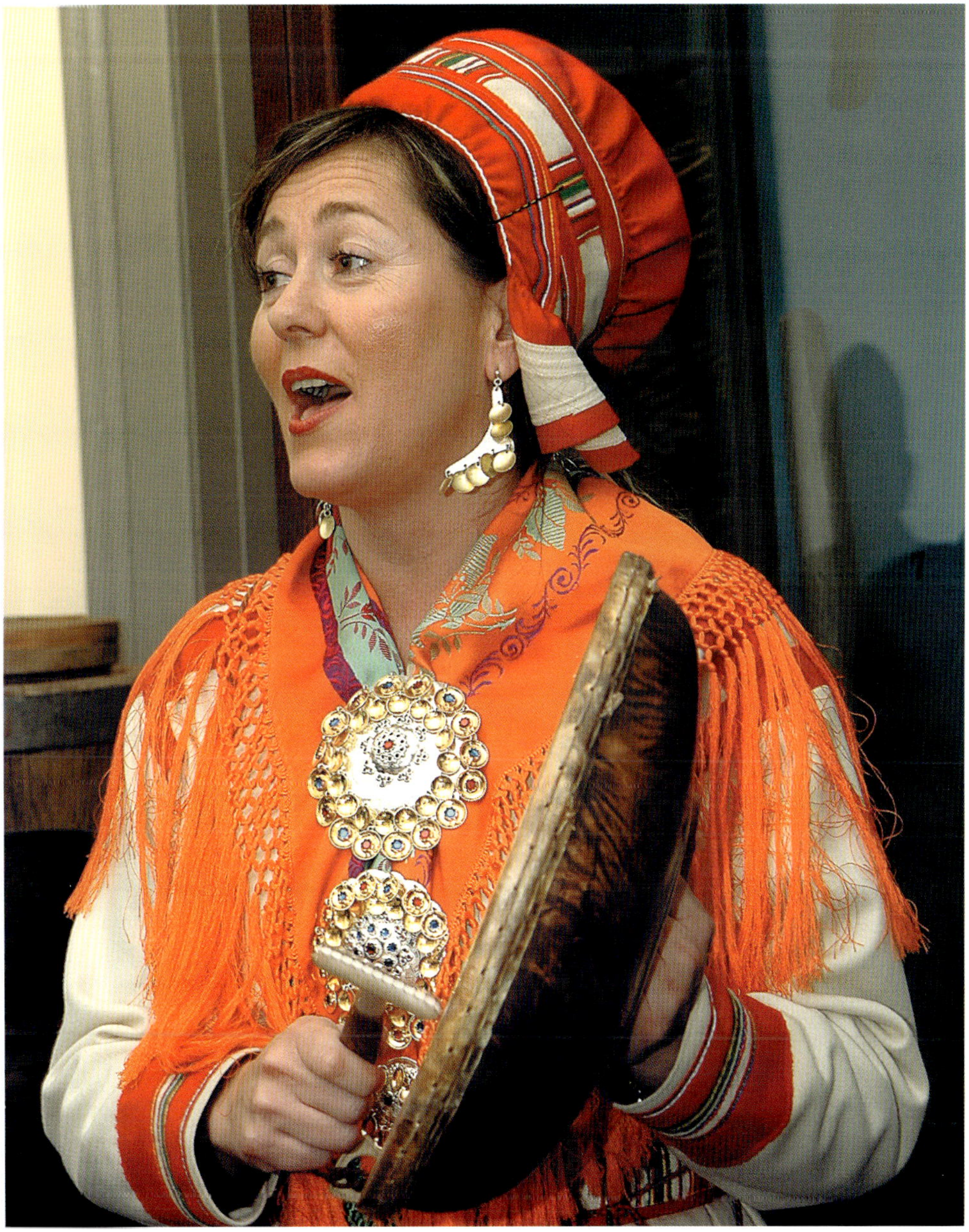

Marit Hætta Øverli performing a Sami *joik* at the *Little Earth* twinning ceremony, West Highland Museum, Fort William,
2 October 2004.
Photo: Alex Gillespie

The story of *Little Earth* Part One was told through the speeches. Each contributor spoke of their ambition for the projects – Alta Museum, the Highland Council, the John Muir Trust, London Fieldworks and the Royal Meteorological Society. It was as if *Little Earth* had galvanised them into action and that the very act of coming together would return them to their agendas with renewed energy.

> "John Muir, after whom our trust is named, is really the pioneer, the father figure of the modern conservationist movement, and he argued for a holistic approach in the management of wild and beautiful places. That holistic approach is not just concerned with the natural heritage, the natural world, but the cultural, the scientific, the humanistic, and the relationship of people with the land. Those ideas come together at the twinning ceremony here today." Nigel Hawkins (John Muir Trust)

Twinning is a common occurrence in civic contexts. Making a particular link between two towns or cities in different countries or continents has proved to be a successful way of extending friendship, bringing together distant communities, opening up trade, healing old wounds, and marking particular historic, geographic or political moments. When the business of twinning becomes an integral part of an art project, it is difficult to distinguish at what point the civic ceremony begins and the artwork ends. Several artists use ceremony in their work – especially in performance where participants would usually be playing a role determined by the artwork's agenda. In Fort William the collaborators brought their civic and professional responsibilities with them. In all their multifarious roles these personalities played a necessary part in the delivering of the work. London Fieldworks invited contributions from a great variety of disciplines – their presence stretching and extending the work, moving it out of an exclusively art domain and into a territory where unexpected partnerships can occur.

The Twinning Agreement signed by Nigel Hawkins (John Muir Trust), Jan Pettersen (Alta Museum), Councillor Olwyn Macdonald, Bruce Gilchrist and Jo Joelson is both the symbol and the evidence of this event. The artwork is defined by it and by everything described above. Its ability to move across national borders and between disciplines helps to establish a more flexible position for art practice. Much in the way that Jeremy Deller's interest in the vernacular resulted in hundreds of local people taking part in a community parade as his contribution to Manifesta 5 in San Sebastian, similarly London Fieldworks, with their interest in the history of two scientific sites, have created a project that cannot be easily delineated. The opening-up of a space for other professionals to fill instils in the project a sense of active engagement – not in the sense of community togetherness but as distinctive, subjective components which transform the work in to a radical monument for the twenty-first century.

Monuments have historically marked exceptional achievement or tragic loss. Explorers create their own informal, often ephemeral, monuments at the climax of their expedition, marking the site with a footprint, a flag or a personal memento. Thomas Hirschhorn gives new meaning to the concept of monument in works such as Monument to Bataille (Kassel, 2002). This monument offered a range of services (taxis, web-cam, café, library, exhibition, outdoor sculpture, local TV studio) and employed 26 local people for the 100-day duration of the *Documenta* show. Just in the way that Hirschhorn unpicks the fixed, solid and immovable notion of monument, so London Fieldworks make connections across the scientific and historical divide, mark the fleeting moment of two cultures embracing, and reveal something special in the everyday comings and goings of a small Scottish town. Twinning *Little Earth* proposes several monuments existing simultaneously in different countries, within different institutions and across different disciplines.

Notes:

1. The Cloud Chamber was imported from Glasgow University's Department of Physics especially for the twinning event.

2. Alta Museum functions as the keeper of the Haldde Observatory.

3. The name 'Haldde' is a Sami name that means 'the Holy Mountain' and it was known as a pre-Christian Sami religion.

4. The John Muir Trust and the Nevis Partnership are the guardians of Ben Nevis and the Ben Nevis Observatory ruins.

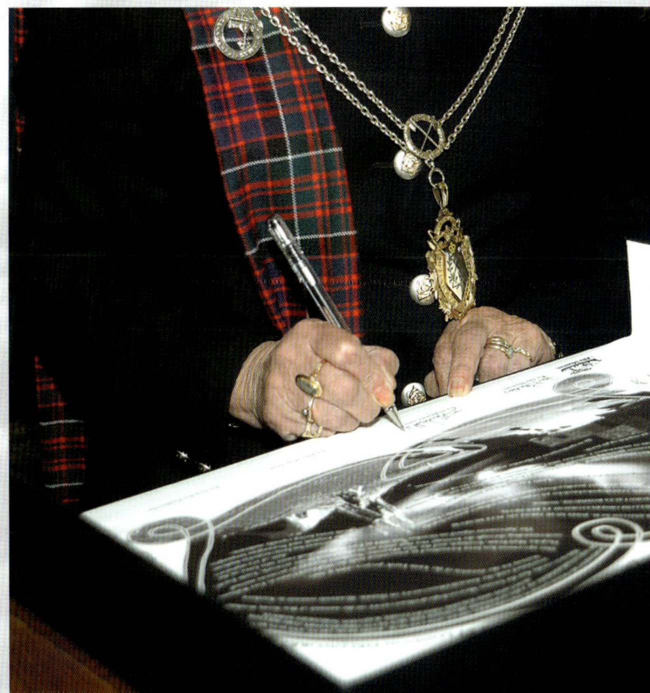

Cllr Olwyn Macdonald JP signing the *Little Earth* Twinning Agreement, West Highland Museum, 2 October 2004. Photo: Alex Gillespie

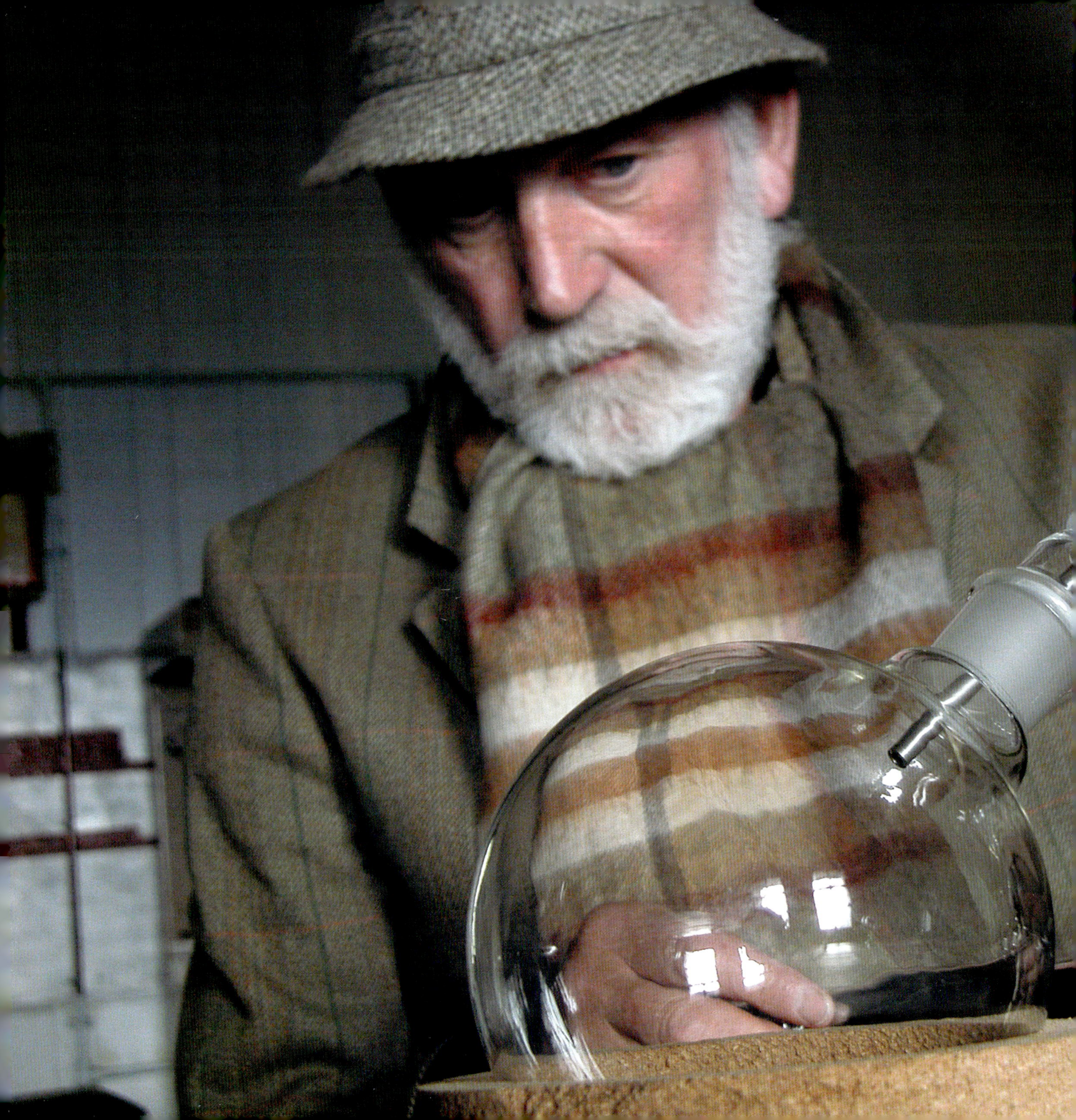

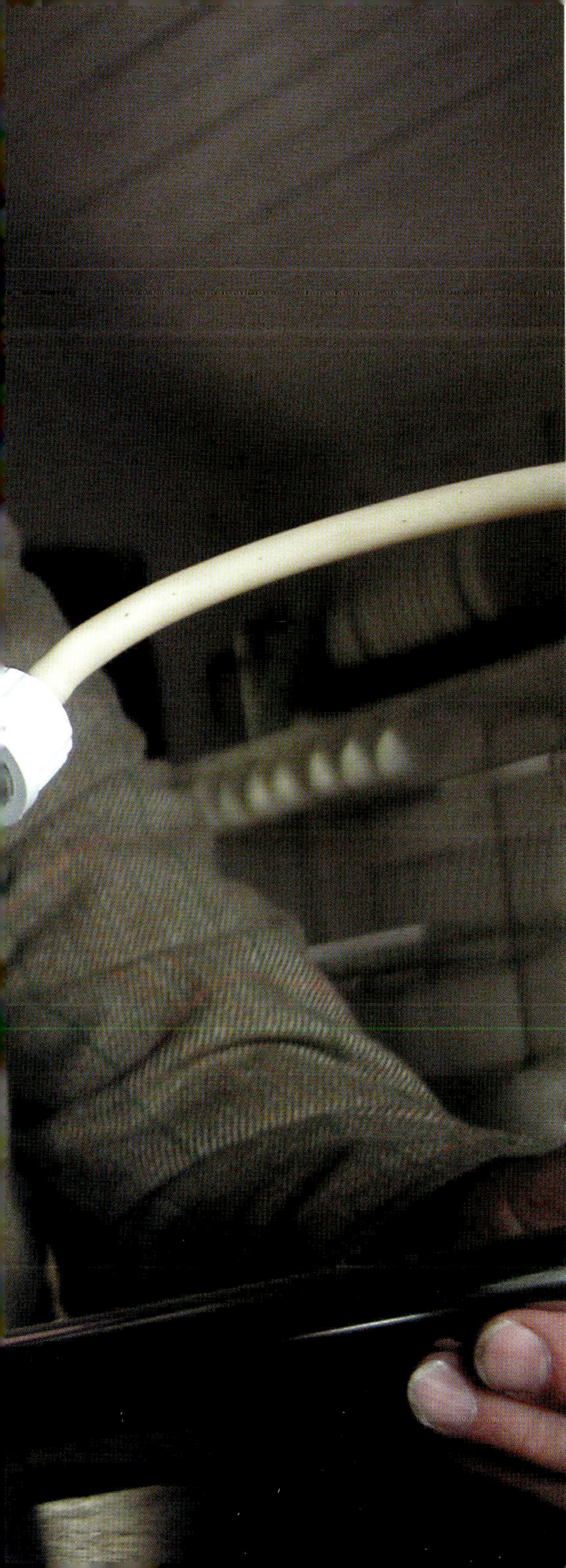

Installation

BRUCE GILCHRIST & JO JOELSON

***Little Earth* production still**
Ben Nevis, 2004
Photo: Alex Gillespie

On 14 January 2005 the *Little Earth* exhibition opened to the public at the Wapping Hydraulic Power Station. On the same day the spacecraft Cassini-Huygens relayed data as it dropped the Huygens probe on to the surface of Titan, one of Saturn's 31 moons. Titan is the only moon in the solar system with an atmosphere. This atmosphere also resembles that of the early Earth and study of its chemistry will provide unique data on the origins of our own planet. The initial launch of the Cassini-Huygens spacecraft in 1997 was controversial: its plutonium fuel caused concern to certain scientists and enraged environmentalists. However, those in the pursuit of scientific knowledge will argue its deployment will pay back in spades.

Little Earth is a project born from the tensions of coincidence, a conjunction of research materials from two previous projects: *Syzygy* and *Polaria*. *Syzygy* was a week-long telemetric art project transmitting body, mind and weather data from Sanda island, off the west coast of Scotland, to the Institute of Contemporary Arts in London. Research for this project introduced us to the story of the former Ben Nevis Observatory and CTR Wilson's fateful connection with it. *Polaria* involved an intensive month-long period of fieldwork camping out on the high arctic tundra. We measured and recorded mind, body and light data, this time translating the recordings into an interface for an interactive installation. Preparing for the field trip to northeast Greenland, a Norwegian artist friend from Alta alerted us to the story of the Northern Lights Observatory on Haldde Mountain and the amazing story of Kristian Birkeland.

The connections between the two stories were immediately obvious but it took three years to produce a work out of this background material. An artist's residency at The Headlands Center for The Arts, (summer 2002) provided access to scientists working at the Space Sciences Lab, University of California, Berkeley, and physicists working at the Exploratorium, San Francisco. Conversations with scientists during this time concerning the significance of Wilson and Birkeland provided further impetus to develop the notion of twinning.

On returning to the UK a relationship was forged with Professor Stanley WH Cowley at the University of Leicester and an arts-science fellowship followed. This provided an opportunity to work with members of the Radio and Space Plasma Physics Group and carry out a visual survey of the progress that has been made in plasma physics from Birkeland's early experiments to the off-world missions currently in operation today. The study began with a review of Birkeland's historic work; from his naked-eye observations documented in log books at Haldde observatory, to the use of instruments such as magnetometers; the development of photographic equipment to capture aurora, and finally the construction of his terrella machine. Much of this work and other research was undertaken via books and articles, through visits to the Auroral Observatory in Tromsø, the Haldde Observatory in northern Norway, the Eiscat Radar at Tromsø, The SPEAR Facility and the Eiscat Radar on the island of Svalbard.

The stories of both observatories, the scientists and their instruments have been well documented and it was never our intention to make a complete retelling of these stories. The main purpose was to develop our perceived connections between the narratives to create a work of fiction drawing on the rich historical sources to create a dialogue between people, places and events.

The official twinning of the observatories in October 2004 served to formalise a number of creative conversations we had instigated in Alta and Fort William over two years. This event transported us across the borderland between imagination and the physical world in much the same way that the cloud chamber and terrella did for Wilson and Birkeland.

The event at the West Highland Museum provided the opportunity to introduce an original cloud chamber to Fort William, the place of its conception. An instrument was disinterred from a dusty basement cupboard in Glasgow University by Dr Alec MacKinnon, ferried to the museum for the day and demonstrated by Dr Andrew Conway. Early in the project development we had considered a resurrection of one of Birkeland's terrella machines in order to include the footage within the video work and we travelled to meet Professor Truls Hansen at the Auroral Observatory, Tromsø to look at an early original. The possibility of rebuilding the terrella machine seemed remote. This led us to consider dealing with notions of machines and instruments in a less literal way, introducing possibilities of image creation and projection for the animation of a sculptural object.

Back at Leicester University the RSPP Group helped to develop and build a representative model for the purpose of filming and inclusion within the video work. Machine I was an early prototype and had a simple rod rotating on an axis with variable speed. Attached to this was a small globe of the Earth containing a magnet. Machine II was a far more complex mechanism, inspired by the solar-planetary relationship and early visualisation techniques. It hybridised aspects of the instruments conceived by Wilson and Birkeland. The machine loosely represented the Earth's orbit with respect to the Sun, including elements such as inclined-rotation axis, variable orbital speed and precession leading to the seasons on the Earth. The main body was a sphere made from a copper ballcock. The machine did not attempt to be a literal translation of these concepts nor did it aim to be an educational tool or scientific instrument, but rather a sculptural interpretation of the Earth in orbit. Most of the parts were hand-machined and constructed by Richard Poyner. To work out some of the problems in building this prototype, an early version was created by another group member, Steve Milan, from his childhood meccano set.

The Cluster mission was to inspire the development of a series of computer animations created at Leicester. Cluster II is part of an international collaboration to investigate the physical connection between the Sun and Earth. Flying in a tetrahedral (triangular pyramid) formation, the four spacecraft collect the most detailed data yet on small-scale changes in near-Earth space and the interaction between the charged particles of the solar wind and Earth's atmosphere. This enables scientists to build a three-dimensional model of the magnetosphere and to understand better the processes taking place inside it.

The four animations have been made from the viewpoint of the craft as they orbit the Earth. Whilst the craft are orbiting the earth the magnetic field lines appear and morph into the most current, up-to-date representative form. The program used to create the animations was initially developed by a previous RSPP researcher, Paul Hamadyk, using data from the 1996 Tsygenenko magnetic field model. Current members of the RSPP group, Jim Wild and Darren Wright, modified this software to create the series of animations for the *Little Earth* project. These animations were made with a view to incorporating them into the larger video work.

For the video shoots on both mountains we were assisted by people familiar with the region. Our guide on Haldde was Arvid Petterson, a northern Norwegian historian with an intimate knowledge both of Finnmark and the observatory. He represented an embodiment of the mountain culture and agreed to perform for the camera. Some months later we found a complementary protagonist for the Ben Nevis shoot in Ian Thomson to whom we were introduced by Alex Gillespie. Ian Thomson had spent many years soaking up mountain experience and possessed the concentration and necessary skills to deliver a performance in far from ideal circumstances.

We filmed both performers in the observatory situation[1] manipulating low-tech objects, representations of proto-instruments emerging from the creative subconscious mind.

The dynamic between the observatory with its outward looking perspective and the encapsulating chamber of an instrument – as if an observatory had been turned in on itself – provided a visual strategy for *Little Earth*. Action was story-boarded, repeat-performed and recorded from the four cardinal points. This allowed us to edit four timelines of video which, for presentation, is synchronised and projected back on to four sides of a projection structure showing multiple views of the same subject. As a poetic echo of Wilson and Birkeland, the record of a perceived event has been re-presented through our 'machine' as a model of spatial reality. Ironically, the exploded view denies one a complete reception.

The relationship to the viewer in London Fieldworks' projects has always been a central concern: in the *Gastarbyter* installation (1998) the work was experienced while being physically supported, reclining supine off the ground; *Polaria* literally electrified the viewer by passing current through the body. Both these works produced immersive experience. For *Little Earth* we wanted to challenge the immersivity of the previous works while continuing to foster non-passive modes of engagement.

In a futile attempt to experience a unification of the projected image in *Little Earth*, the audience may be driven by curiosity to circumnavigate the structure. This introduces a relativity between the image and the audience, inducing a rotation. The observer becomes a particle to a nucleus, a satellite to a planet.

Notes:

1. To stand-in for the former Ben Nevis Observatory interior we shot video sequences in the CIC hut which is owned by the Scottish Mountaineering Club, originally opened in 1929.

Little Earth video stills,
London Fieldworks 2004

Script

JAMES FLINT

1. INCOGNITOGATEN

Voice of CTR: What's the last thing you remember?

Voice of KB: The last thing I remember? The house on Incognitogaten, the boat to Egypt, a garden in Japan.

Voice of CTR: Places only? Nothing more specific? Nothing else?

Voice of KB: Some things, some… A gun, a bottle of Veronal, the pillars of Hercules, the golden path the Pharaohs trod to take their place among the stars.

Voice of CTR: That would be… the Zodiacal light.

Voice of KB: No – something else. A blazing trail, vertical – the fire tail of a powerful rocket-ship, climbing out of somewhere near the Caucusus, up and up until it burst free from the atmosphere and arrived in outer space.

Voice of CTR: And now you're one of us.

Voice of KB: You? Who are you?

Voice of CTR: We're the eyes that see. The sprites that watch the storm.

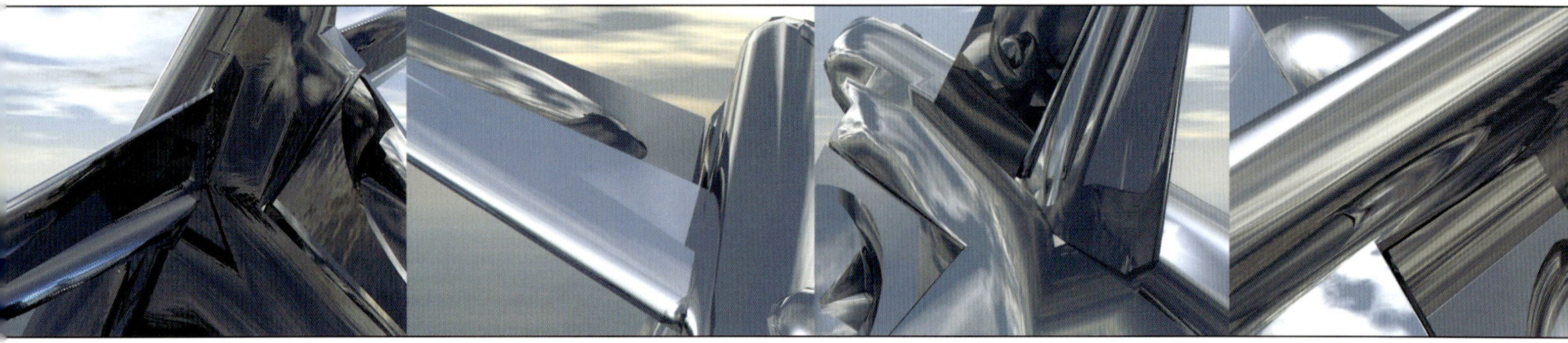

2. RETURN TO BEN NEVIS

Voice of CTR: I was on a ship, a flying ship, but a strange ship – no hull, only a mast, and sails – great metal sails, long, thin, rectangular. A ship that tacked across the black silent ocean of the sky, a ship that was measuring, always measuring. Beneath me was a mountain, its peak enshrouded with a scarf of mist.

Voice of KB: Beinn-neamh-bhathais [old Gaelic name for Ben Nevis].

Voice of CTR: And then it came to me. I was here before. An image in the mist, Leonardo's Vitruvian man, that's what I remember. Hands and arms like vessels charting a circumference. A garden of rocks and lichen alive with wee snow buntings. And icicles choking the louvres on the instrument screens like so many flying spears.

Voice of CTR: As I watched the mists began to move, not gently, as before, but violently. The clouds solidified into a series of peculiar, shifting forms, as if trying to emulate the structure of thinking itself. And then, the mists cleared and I found that I was gazing down at the tiny simulacrum of a room, a room I recognised from photographs even though I had never set foot in it myself.

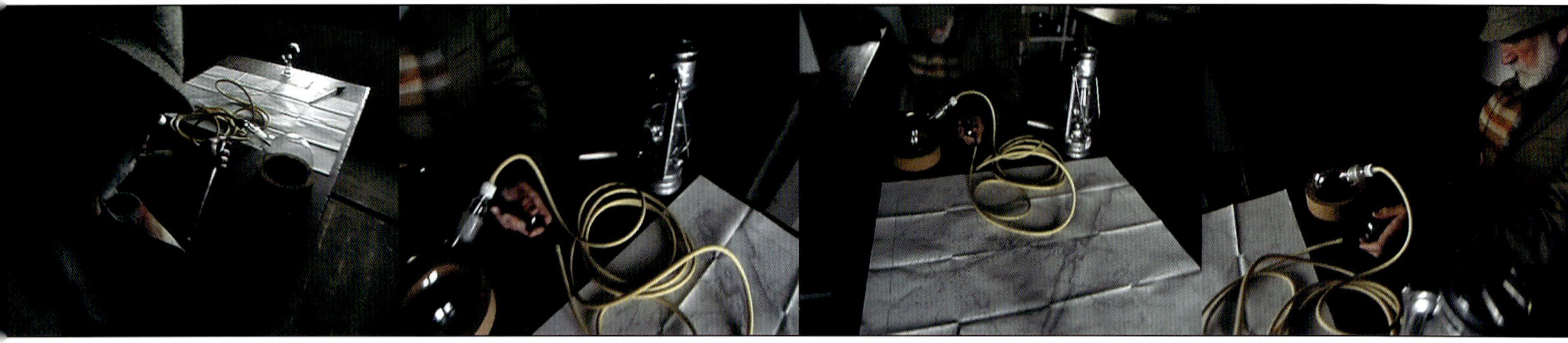

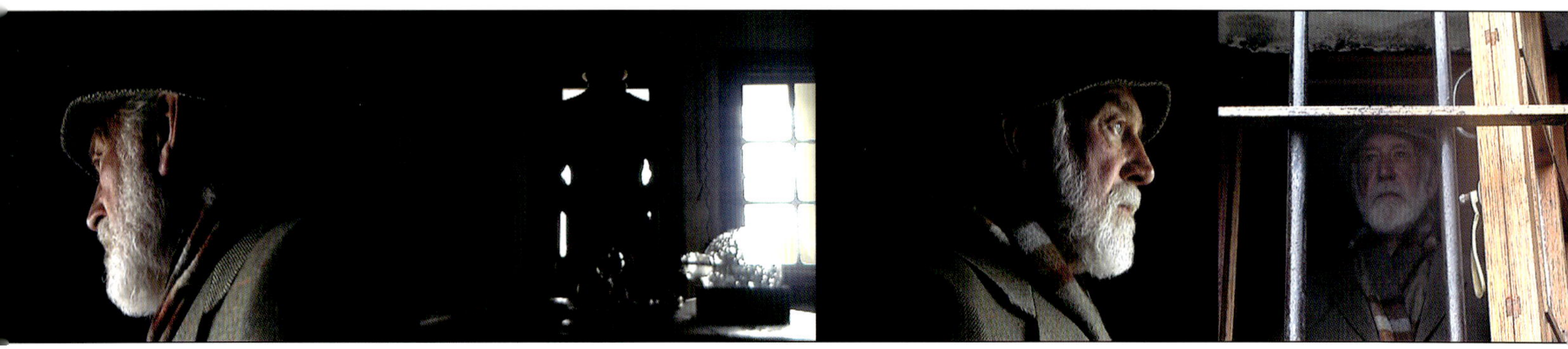

3. THE EXPERIMENT – BEN NEVIS OBSERVATORY

Voice of KB: What room was it?

Voice of CTR: The room was that which had served as living quarters in the observatory, empty save for a table and a single chair.

Voice of KB: I know the place.

Voice of CTR: How was I to know that when I pumped the gas from the vacuum flask I'd seed a whole new universe? That I'd see electrons streaming down through the atmosphere, raise the curtain on the play of matter itself? Who would have known that all that could be found in clouds, in simple vapour trails? Who could have guessed that from this nest of particle activity, from this adiabatic egg, a demon would emerge, a sprite, a raiser of temples, a destroyer of worlds?

Voice of KB: And if, perhaps, you could cancel the spell? Push the genie back inside the bottle? Retrace your steps, reverse the process, unwind the skein of your own history, and erase the tracks you laid around those old stone walls?

Voice of CTR: But nothing can be undone. The journey made cannot be untravelled, the path chosen cannot be changed. There's nothing for it but to go on and on, round and round and round again, to spin and spin and overlay your self, the Earth, your past, your memory with so many tracks, so many passages, so many particles that everything becomes indistinct and a vortex is created, a vortex that sucks in the atoms of existence and blurs them in a dervish whirl, flattened and immane. Terrible, but beautiful.

Voice of KB: And beauty is truth, said Keats.

Voice of CTR: Keats was wrong. Beauty is information.

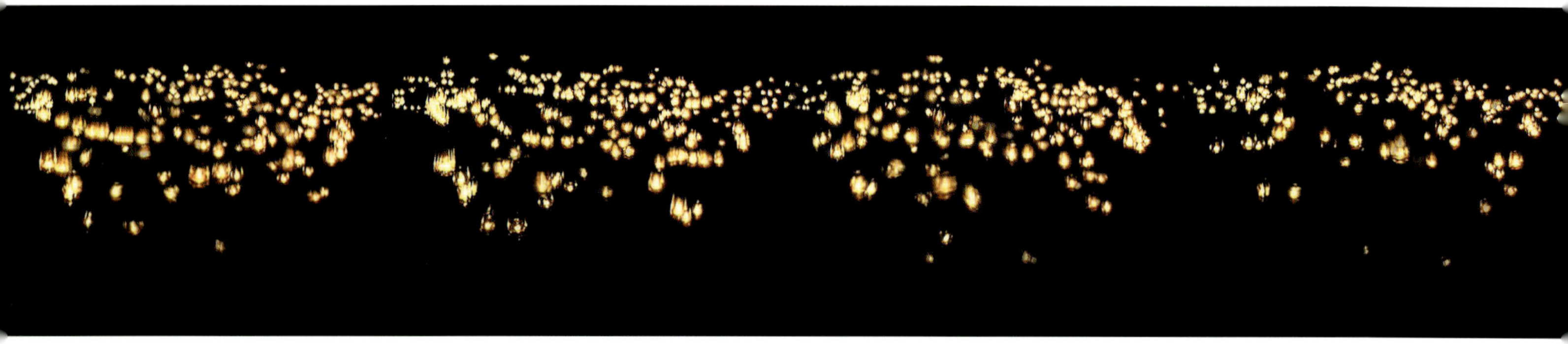

4. THE STORMLAMPS

Voice of KB: There it was, floating before him, a plane in the night–

Voice of CTR: The cinema of the very small–

Voice of KB: Filled with stars–

Voice of CTR: The fame of matter–

Voice of KB: Quantum glamour–

Voice of CTR: The wonder of the world conjured into view–

Voice of KB: By our mediums, the machines.

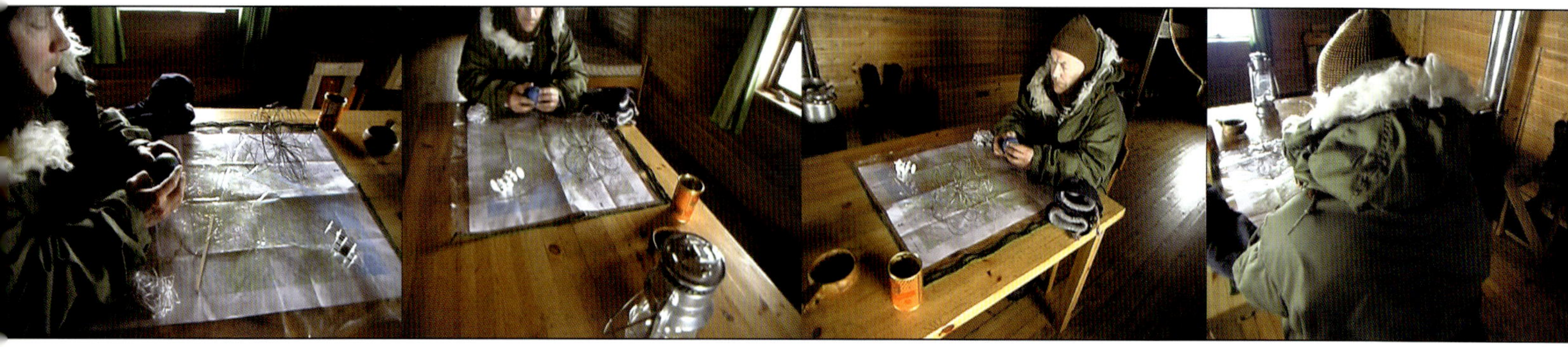

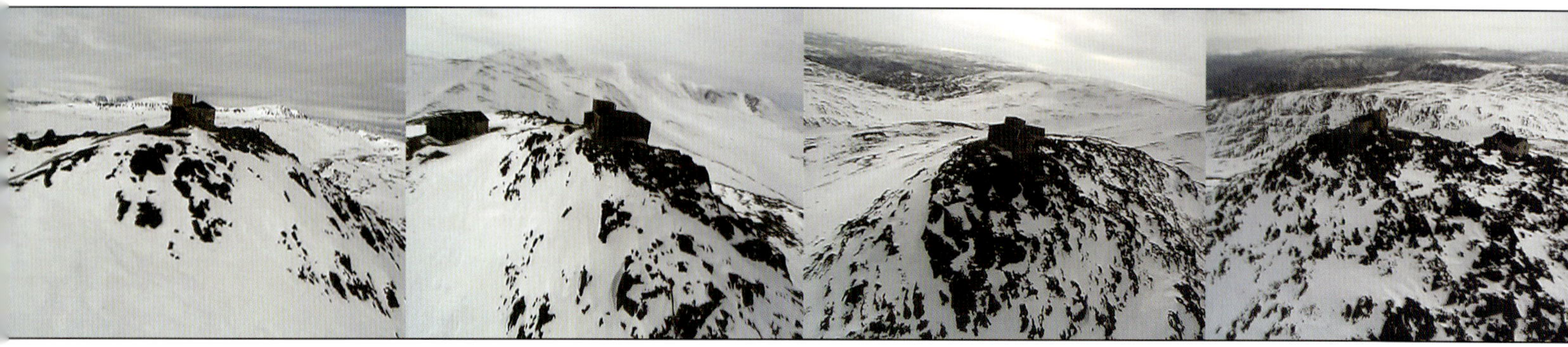

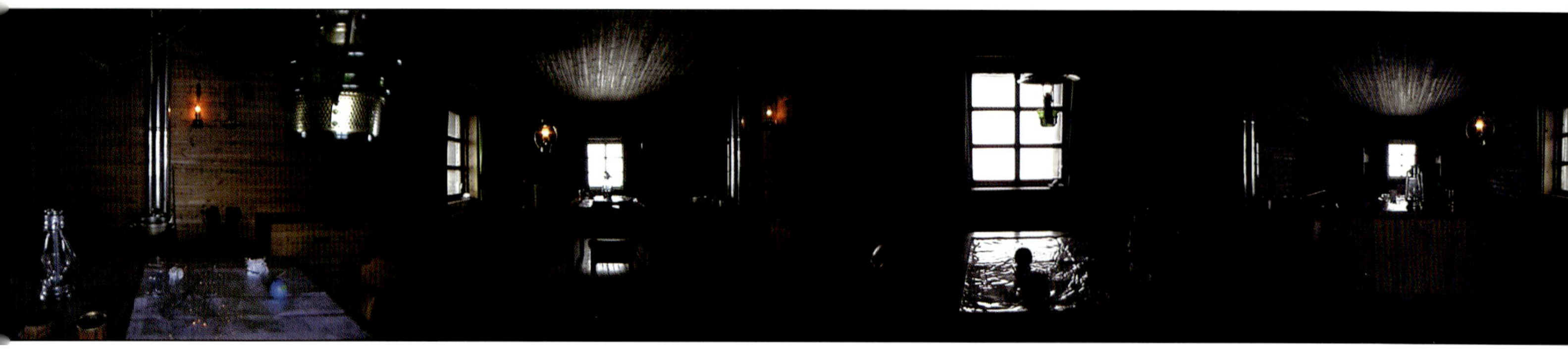

5. RETURN TO HALDDE

Voice of CTR: It's over for us now, Kristian. We're the last who'll be able to ever see things on their own. The human eye–

Voice of KB: The human eye is no longer set inside the human skull. And I, for one, find that something of a relief.

6. THE EXPERIMENT – HALDDE MOUNTAIN OBSERVATORY

Voice of KB: I was in orbit. I was out there looking down at the globe of the Earth and people were the thing most invisible to me.

Voice of CTR: And then – someone called you and you came.

Voice of KB: I came but there was no one here. No one, nothing, no instruments. This is not the past but a vision of it. This isn't the past, it's a simulacrum. I don't remember much, but I do remember I was never on my own here.

And here, this thing, what's this?

Voice of KB: It's an artificial eye. The human eye sees spheres of light and harmony with darkness on the outside. This eye lets you see the reality of things. It's an eye and finger, both. It lets you touch the fabric of the universe.

Voice of CTR: A lodestone disguised as a little Earth.
What were you looking for? What did you want?

Voice of KB: I wanted to stand on the mountaintop

Voice of CTR: You wanted to disappear?

Voice of KB: I wanted to go into the light.

Voice of KB: I wanted to become part of the mechanism.

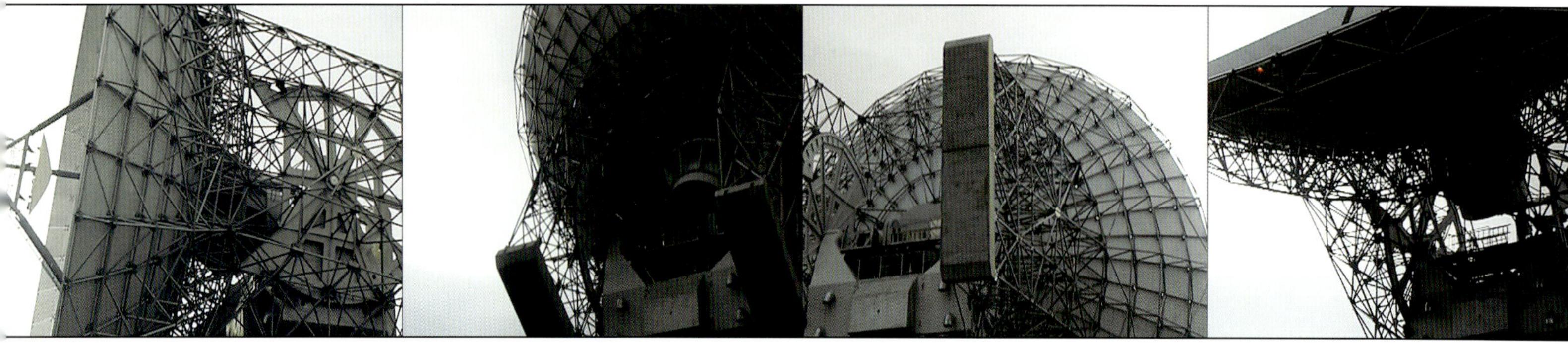

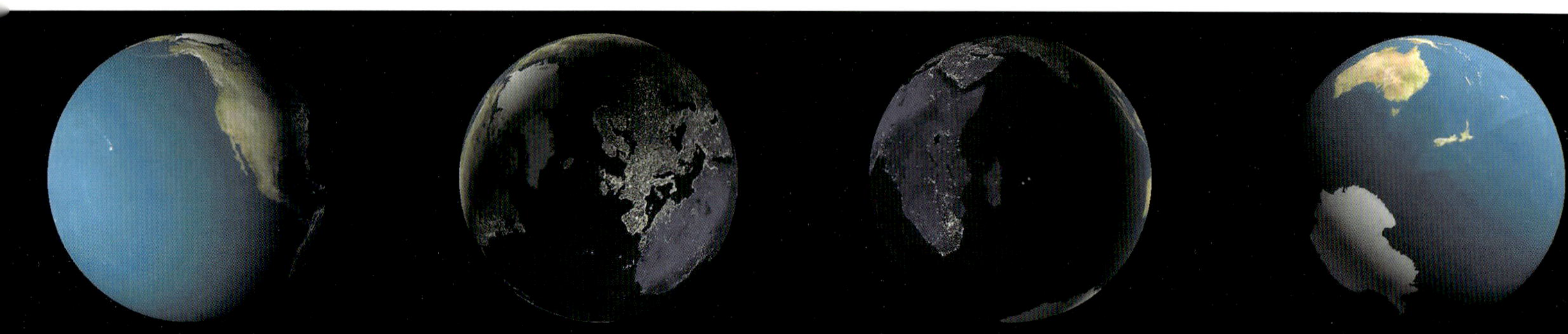

7. ORBITERS

Voice of CTR:	There are four of us, matched in space, a tetrahedral pyramid.
Voice of KB:	A structure to focus power, to send a laser light between the heavens and the earth.
Voice of CTR:	An eye to see.
Voice of KB:	A telescope for looking upwards, forwards, outwards.
Voice of CTR:	A surveillance tool for looking inward, downward, back.
Voice of KB:	A probe that hitched a ride around the planet.
Voice of CTR:	Bound for Saturn, the sphere without a surface, little Sun.
Voice of KB:	Bound for Titan, ancient Earth, moon of nitrogen and methane smog.
Voice of CTR:	A chance for a plutonium powercell to vaporise and seep into the atmosphere.
Voice of KB:	A chance to reach out into the velvet night and touch another world.
Voice of Ida:	Connect.
	And interface.

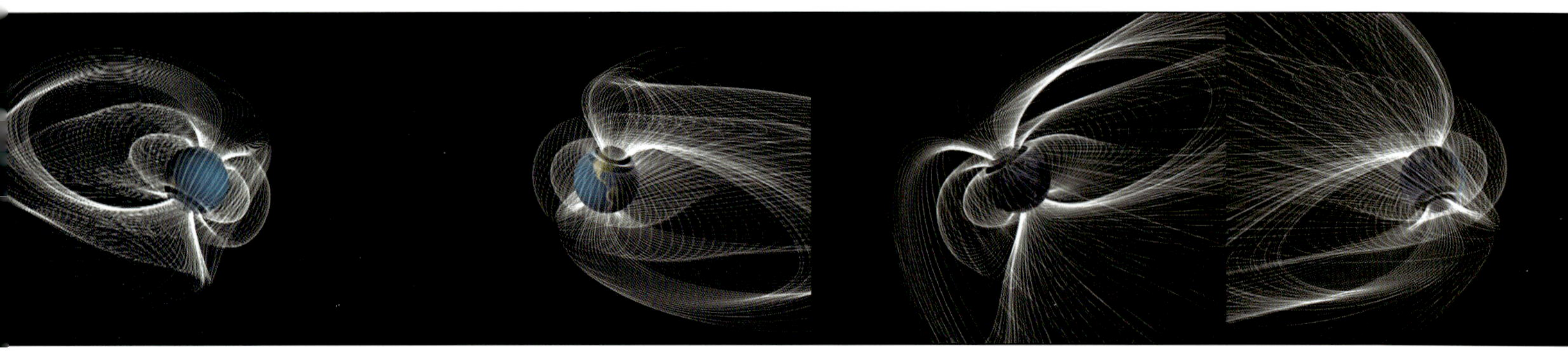

8. THE VISION – CLUSTER

Voice of Ida: Fluxgate Magnetometer... Electron Drift Instrument... Active Spacecraft Potential Control Experiment... Spatio-Temporal Analysis of Field Fluctuation Experiment... Electric Field and Wave Experiment...Waves of High Frequency Sound for Probing of Electron Density by Relaxation Experiment... Research with Adaptive Particle Imaging Detectors... Plasma Electron and Current Experiment... ASPOC... STAFF... WHISPER... RAPID... PEACE... ASPOC... STAFF... WHISPER... RAPID... PEACE... RAPID... PEACE... PEACE...RAPID... PEACE... PEACE... PEACE... PEACE...

Voice of KB: Ascended from our mountaintops, sacred mountains, having dethroned their ancient gods. A dream of light and clarity from a world that's all but gone.

Voice of CTR: We hover at the corners of the planet, a trap for particles flying in at five hundred miles a second. From watching trails of light inside a vacuum chamber, we've become trails of light across the sky.

Voice of KB: Ideas, mere reflections of man, placed behind the majesty of the everyday operations of matter. It's war, always war. The solar storms rip at our magnetic field, the earth a wind-driven ship in space. Our iron hearts alone – that's all that saves us being torn to shreds and scattered throughout the corners of the cosmos.
The magnetosphere.

9. RETURN TO INCOGNITOGATEN

Voice of CTR: I am here no longer; this mountain, the town below, all gone. Nothing but a flux of energy whose pull and gush fills the lochs of time, a cosmos of greedy whirlpools of which this matter is the merest spurl. Walker, thinker, dreamer, that's the single destination, that's the life we live, that's the grave we're going to. This nature, the ruin of our science; these jagged walls, these fragments of our folly; these sky-ships, these missiles, these shards of sun held in chains on earth; these broken teeth with which we strove to chew our way through everything: this trembling ground, this sheltering sky, this little Earth.

Little Earth: Sound, Cinema, Sci-Fi, Space

1. Sound and Cinema

In asking 'what does cinema want of sound?' the answer is usually that it wants realism – reality from a human perspective – from pre-history to imagined futures to parallel universes, regardless of genre or style. In practice, this means: Foley to provide indices of physical materiality, the hiss of a sword or the charged hum of a light sabre; music, as soundtrack, to provide pure ringing tones of sincerity, ensuring that what is seen is also felt, emotionalising -scapes, amplifying the shape or effect on the faces of innumerable actors; clarity of voice, the absolute signifier of presence and governor of narrative vector; and, above all, unity, wholeness, the alchemy by which sound temporalises cuts between shots to create scenes (the illusion of linearity) and overarches disjunct scenes, stitching time together with musical leitmotifs and sonic themes. Sound gives cinema 'the wholeness of its illusion' (Cubitt, 169). In *Little Earth*, which references but centripetally explodes cinematic reality on to four screens and 8 speakers (which are not anchored to the vanishing point of a central speaker behind a single screen), the conventional relationship between sound and image is constantly expanding and contracting, stretching towards abstraction, snapping back to provide indices of materiality and emotion. The constant in this is sound's role as provider of temporal unity, invisibly linearising the cuts that separate shots, or reinforcing these in moments of 1:1 synchronicity – 'perceptual solidarity' (Chion, 37) – between sound and image.

Sound is the essential element in film's formula – 'unification of multiplicity under the sign of mortality' (Cubitt, 169). This is a possible definition of classical cinema, which 'has only the wholeness of its illusion' found in 'the hard edge of referential reality, the adamant certainty of the image' (230). By contrast, neo-baroque cinema – typified by sci-fi films such as *Armageddon* and *The Matrix* – 'tends towards the cloudy and disorientating... resulting in a preference for questions to be left unanswered, identities to be guessed at' (230). In *Little Earth*, sound plays a double role, sometimes corresponding to its use in classical film – providing indices of the everyday – and at others corresponding to the neo-baroque – adding value to the ordinary, transforming it into giddying moments of scientific epiphany, a whisper of what is waiting to be discovered just beyond the reach of human observation: the globe in Birkeland's hands resounding magnetically; Wilson's vacuum flask breathing the secrets of particle energy. Even as such moments pull the receiver into fogs of metaphor, there remains the concrete physicality of sound which, while pointing beyond the observable, also reminds us that at some level this is all real; a reality that can't be touched or seen, but which touches us, tickling the auditory filaments, echoing in the chambers of the gut. Reality, both cloudy and hard-edged, is always audible even if not decodable.

In normative cinema, the balance between the grain of sound, its intrinsic properties, and the semantics of sound, its extrinsic referentiality, is tipped sharply towards linguistic meaning, towards verbocentricity (Chion, 5), against which the grain of the voice only infrequently rubs. This is the logic that underscores sound in film: the directive that sound is to support what is seen – excepting moments of madness or tragedy where lack of empathy between sound and image sharpens the blade of these deviations. Loosening the bonds of this convention allows exploration of the oft proclaimed but rarely utilised potential of sound to diffuse concentric rings of meaning, radiating outwards from 'common sense' in ever fainter harmonics which, despite their apparent dissonance, continue to bear relationship to the semantic fundamental.

There are of course many astounding examples of such remote and telling audio-visual dyads: the saccharine pop innocence that heightens abjection in the films of David Lynch ('...she wore blue velvet...'); the waltz, Strauss's *An der schönen blauen Donau*, that lends absurd grace to the space station in *2001*; the breaking glass that freezingly signifies the suicide by liquid oxygen of the heroine in Tarkovsky's *Solaris*. In *Little Earth*, sound shuttles back and forth between the normative, where there is a perceived audiovisual consonance (a match is struck and the sound of a match strike is 'spotted' to the image), and dissonance in which sound and image have no conventional relationship – swarming storm-lamps whorl about the screens before singularising into a Highland reel. Likewise, the voice is treated not simply as a carrier of linguistic meaning, but also as a fundamental pitch from which more remote harmonics can be generated; hence voices are literally atomised, granulated into particles that are rearranged into other forms, the crackle of ice crystals, the bursting of water droplets, the bark of static. This, as Wilson puts it in *Little Earth*, is the 'vortex that sucks in the atoms of existence and blurs them in a dervish whirl'.

2. Sound and Sci-Fi

Sci-fi, the height of the filmic neo-baroque, floats in the allegorical space between what is portrayed and what this can be taken to mean (Cubitt, 230). This well describes *Little Earth*. When Wilson activates a cloud chamber or Birkeland contemplates the globe, we witness acts of observation, interpretation, attempts to bridge the space between phenomenological and scientific reality, a gap that can be understood as allegorical. When the normative 1:1 logic of film is abandoned, the possibility of allegory – of metaphor – is also reached. Wilson enters his hut on Ben Nevis, we see and hear the door open. He closes the door and we see this, but we do not hear the door close. In science, a door that has been opened cannot again be closed (Wilson, 'Nothing can be undone'). Rather, it leads to other doors, which once arrived at must be opened. The more doors opened, the further science is from phenomenology, from the veracity of the naked eye's observations. In *Little Earth* there is a Pandora's box of such portals, refracting and reverberating in the sound space of the installation.

The more doors science opens, the more it requires devices – prosthetics – to see hear touch with (Birkeland, 'Our mediums, the machines'). The more the reach of human observation is superseded, the more technology is imbued with superhuman power, taking on a godlike aura that is most evident in the totemistic reverence shown for devices capable of great speed (both micro- and macro-) and/or of navigating impossible distances (from nano- to cosmic). In *Little Earth*, sound responds to the human/machine dialectic by looping wetware within hardware within software, fabricating a soundscape to remind us that the ghost in the machine is ourselves. When devices chatter hum bark at us, they do so with voices in which the grain of the human is still perceptible; and as these objects are human in origin, they – as we – are destined to fail. This is machinic sin, the trajectory of the dying device tracing the arc of the Fall, the honesty of the machine's last confession measured by the ratio of signal to noise (think of HAL in *2001*).

3. Sound and Space

The greatest trick that sound is capable of is the illusion of presence in absence. We see nothing, but our ears tell us that something ought to be there (acousmatic sound). Once associated with divinity and the supernatural (in the beginning was the word and the word was…), the phenomenon of not seeing what is heard is now a technological ubiquity. Still revered by electroacoustic composers, the acousmatic is so commonplace as to be banal – there is not a shopping mall in the world which does not feature 'piped' voices exhorting patrons to celebrate their corporeality through consumption. Yet in cinema, the voice that is heard but not seen still carries a power associated with the superhuman and the spiritual: in *American Beauty*, the voice that ends the film is that of the dead protagonist elevated from the corpse of a suburban rebel (whose voice issues from his mouth) to become the spiritualised omniscient narrator who drifts with the clouds, his voice the only proof of his existence; in *Little Earth*, the voices of Birkeland and Wilson are similarly transmuted: liberated from the scientific struggle to know more they come to know everything, a mystical knowledge born of bodily dissipation; in ceasing to be themselves they become everything and thus their voices merge with and emerge out of the particle substrate.

The magic trick of the acousmatic is somewhat lost in film where the image 'magnetises' sound to particular points in space; footsteps are heard emanating from the figure we see walking on screen even when we watch a film wearing headphones (as on aeroplanes). In this sense, the image anchors sound in space just as sound anchors the image in reality. Yet the power of sound to transform the space of the image remains intact: composed space (that of the soundtrack) opens up the possibility of spatial extension, capable of expanding the frame from inner space – a character's psyche – to the infinite spaces of the off-screen environment. In some instances, these extremes are presented simultaneously, as is the case with the jungle noises that envelop Capt. Benjamin in his Saigon hotel room in *Apocalypse Now*. At other moments, sonic space points beyond the real, as do the battle cries of Broken Sword, Flying Snow and Sky in *Hero* – cries which reverberate in the afterworld (anticipating the demise of these warriors); and there is also real space, the space of surroundsound, under-utilised even in the technologically driven imagination of Hollywood, offering the possibility of physically extending sound far beyond the image, a capability perfectly suited to the vast and empty -scapes of *Little Earth*. Like science, sound cannot be contained, only followed as it echoes into the spaces it opens.

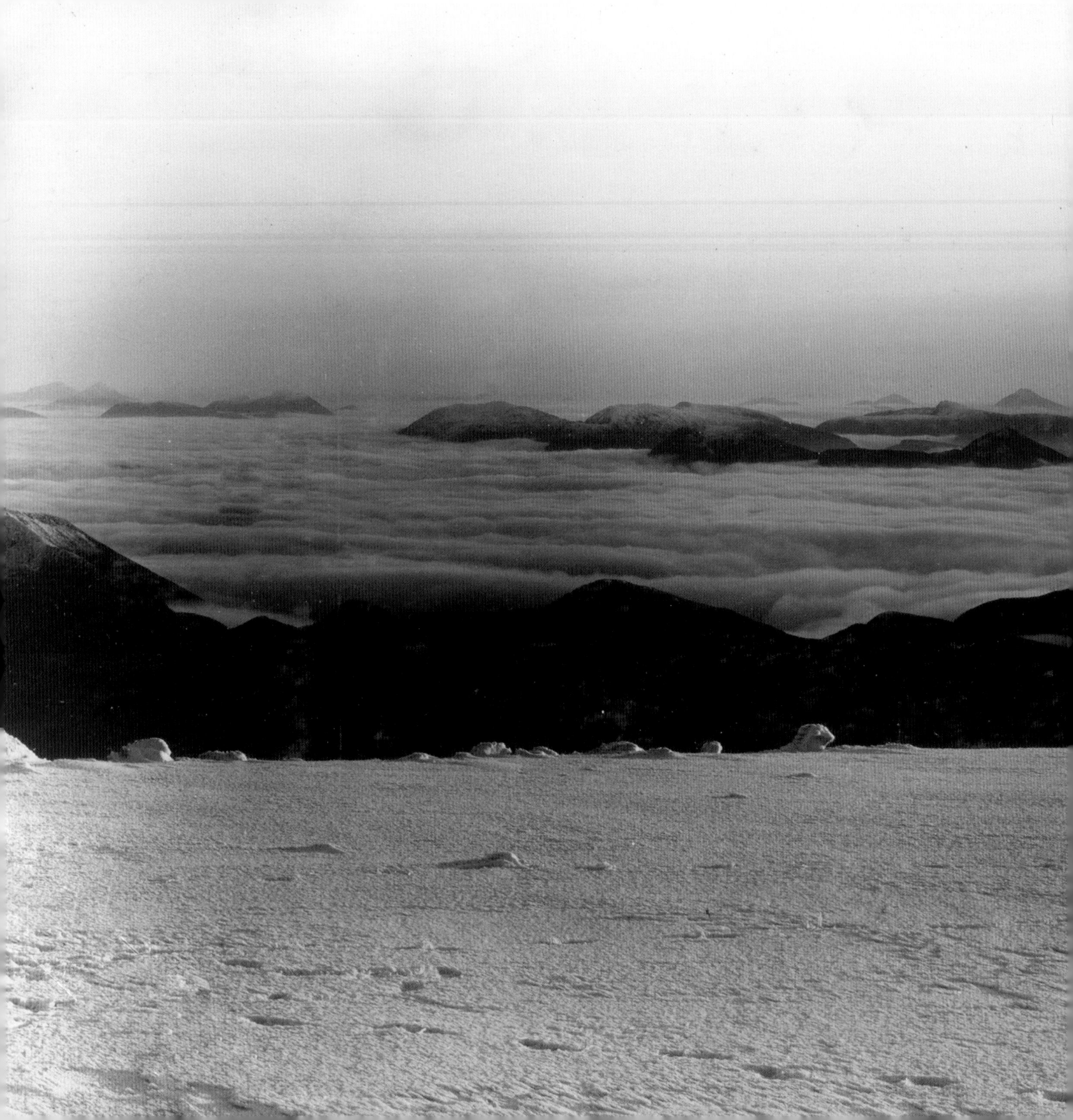

CTR Wilson & The Ben Nevis Observatory

MARJORY ROY

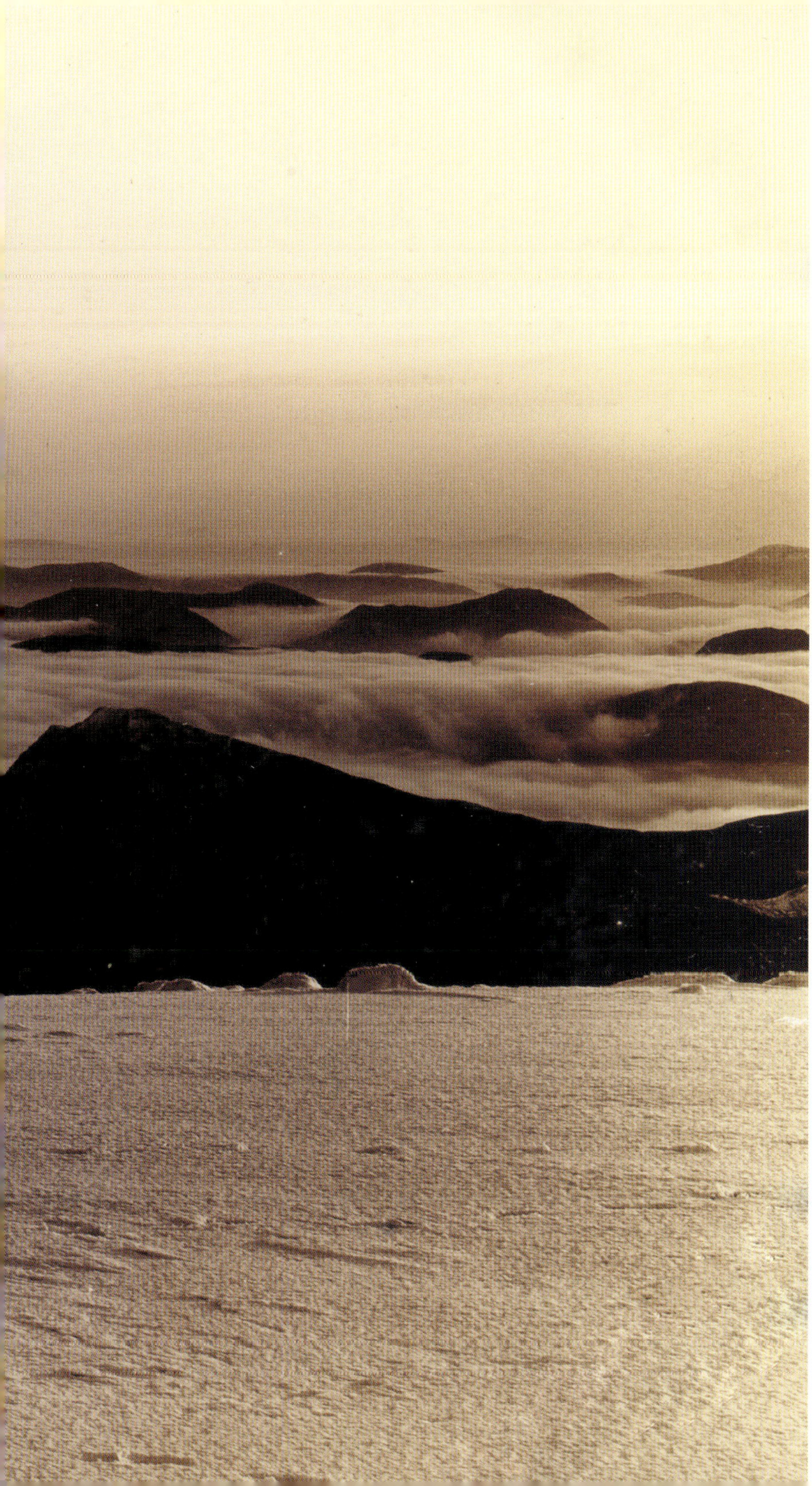

View from Ben Nevis
with a temperature inversion, c.1894
Courtesy of Edinburgh Met Office

WILSON'S BACKGROUND AND EDUCATION

CTR Wilson was born on 14 February 1869 at Crosshouse farm, Glencorse, at the edge of the Pentland Hills, to the south of Edinburgh, but his father died when he was four and the family moved to Manchester. He attended Owen's College (now the University of Manchester), concentrating on biology, intending to become a doctor. In 1888 he went to Sidney Sussex College, Cambridge, with a scholarship and became interested in the physical sciences. He graduated in 1892, having abandoned medicine as a career. In 1894 he heard that young physicists were welcomed at the Ben Nevis Observatory as relief observers and offered his services.

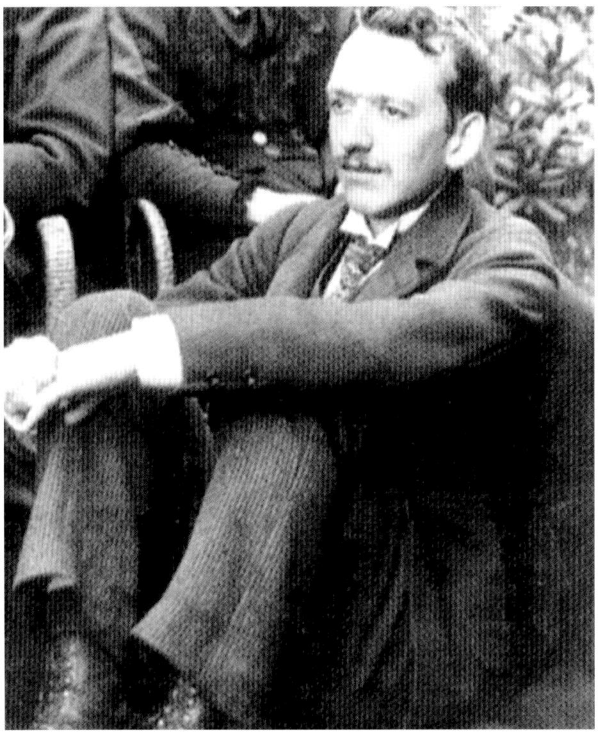

Portrait of Charles Thomson Rees Wilson *(1869-1959)*
Plate 8, scanned from Notes and Records of the Royal Society vol 14, no 1, pp 163-173 (1960)
Courtesy The estate of CTR Wilson

THE BEN NEVIS AND FORT WILLIAM OBSERVATORIES

The Ben Nevis Observatory was set up by the Scottish Meteorological Society (which amalgamated with the Royal Meteorological Society in 1921) in 1883, following a public appeal, in order to study how weather conditions vary with altitude and so improve weather forecasting. Due to the severe environment, especially the heavy icing and strong winds at the summit of the mountain, the observations had to be made manually every hour. This meant that there had to be at least two observers resident there at any one time and there was also a cook. A Low-Level Observatory in Fort William was opened in July 1890; there measurements were made with automatic photographic recording instruments, which were supplied by the Meteorological Office. The number of permanent staff was increased after the opening of the Fort William Observatory, with staff changing over between the two sites. In order to let permanent staff take holidays, students were welcomed as relief observers during their summer vacation.

WEATHER DURING WILSON'S WEEKS AT THE OBSERVATORY IN SEPTEMBER 1894

Wilson's period of duty from 8-22 September 1894, was one of the driest in the 21 years of observations, with high pressure over Scotland throughout. Weak fronts gave a little rain on the night of 10/11 September and the morning of 14 September, but otherwise the period was dry with light winds. The four pages of the weather log covering his tour of duty contain frequent reports of clear skies above and fog or haze trapped in the valleys below, indicating 'inversion' conditions associated with anticyclonic descent of the air. Normally temperature falls with height, but with an inversion the temperature actually rises with height through part of the atmosphere. Above the inversion the air is very dry. Photographs taken from the Observatory under such conditions showed a sea of clouds or fog, with the higher mountains appearing like islands. The inversion level varied throughout the fortnight, being sometimes above the mountaintop, with hill fog covering the peak, sometimes below. On 19 September it was around the level of the summit since 'Temperature was unsteady, rising several degrees when top was clear & falling an equal amount when fog was passing'. The inversion conditions were ideal for the observation of the glories and coronae, which so impressed Wilson and the log for 15 September reported 'Solar fogbow & glories at 16h & lunar corona at 23h'.

CTR Wilson & The Ben Nevis Observatory

MARJORY ROY

In order for a glory (or Brocken Spectre) to be seen, the sun must be behind an observer and shining on mist or cloud in the opposite direction. It is very rarely seen on flat ground, except at the edge of a fog bank at sunrise or sunset and the best situation is looking down from a mountain top on to cloud in a corrie below. (A glory may also be seen around the shadow of an aircraft on a bank of cloud below.) An observer sees his shadow on the cloud, with his head surrounded by one or more coloured rings, which form the glories. If more than one person is present, each sees the glories only around his own head! If the cloud droplets are all of about the same size then several glories may be clearly seen, but if they vary considerably in size the outer rings disappear and the inner one tends to be white. The droplets also need to be of the right size (about one hundredth of a millimetre across) to give well-defined glories.

The coloured glories are formed in the same way as rainbows, only the reflections and refractions of the sun's rays occur within cloud droplets, which are very much smaller than the raindrops which give rise to rainbows. The summit of Ben Nevis was an ideal situation for observing glories, but on average they were only seen about once a month, showing how relatively rarely conditions were ideal for their formation. The early log books contain a considerable number of drawings of the most striking examples, with measurements of the angles of the various rings. Much more commonly seen were coronae (often double or triple) in clouds passing across the moon, and haloes and mock suns in high level ice crystal (cirrus) cloud.

A notable feature of the weather of Ben Nevis was the very high frequency with which cloud covered the summit, giving hill fog. Over the twenty years of operation of the observatory the average frequency of fog in September was 70%, so Wilson was exceptionally lucky in the conditions that he experienced in 1894.

Writing in *Weather*, 60 years after his fortnight on Ben Nevis, on the occasion of the fiftieth anniversary of the closing of the Observatory, Wilson described the conditions that he encountered during his brief sojourn at the apex of Britain's highest mountain. He was responsible for the hourly observations starting at 5 am each day, which meant that, when the sky was clear and there was cloud below, as the sun rose he would see to the west the shadow of the Ben racing across the surface of a sea of cloud. 'On the cloud surface beyond the shadow would then appear a glory, the coloured rings rather faint and diffuse. More brilliant glories were to be seen at other times of the day'. Lunar coronae were also frequently seen during his time at the observatory.

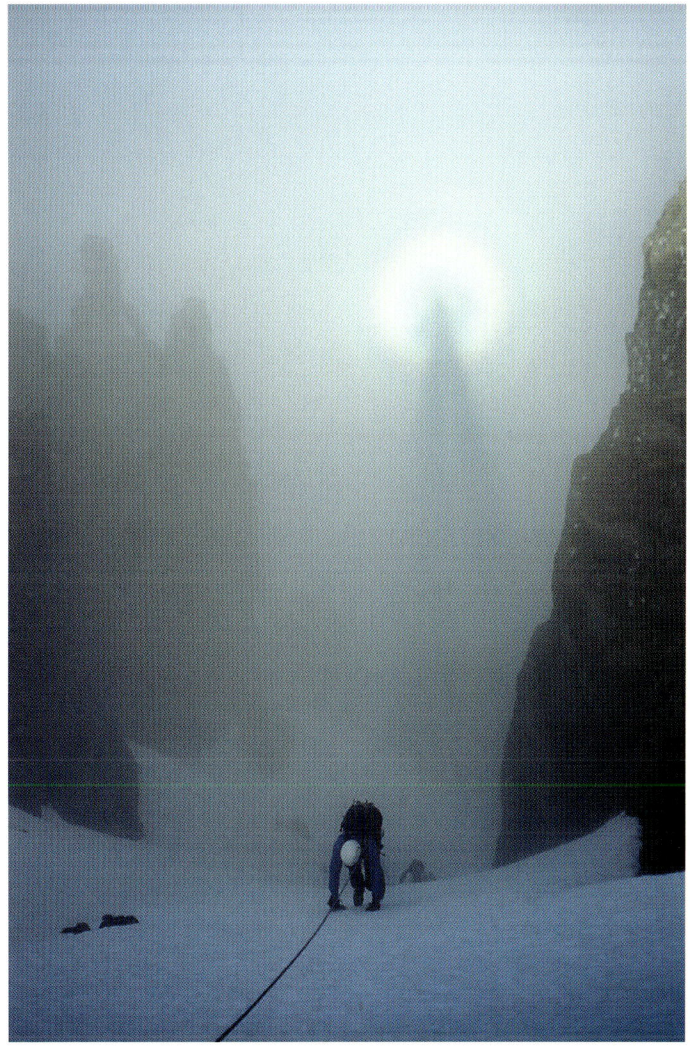

Glory at Hayfork Gully on An Teallach,
February 2003
Photo courtesy of Mick Tighe

Similar corona seen at Mdnt. Colours as in diagram, the outer yellow a very narrow strip. Occasionally traces of a third ring were observed.

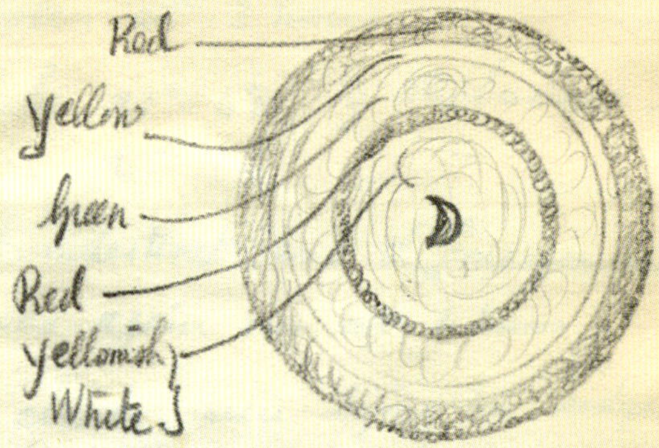

Radius of outer red 4°. 56'.
" Inner red 2. 11.
" ~~inner~~ yellow 4. 23.
} all good measurements.

October 23.

Double corona at 2 hrs, with at times faint third ring – colours as in diagram – No blue or green inside the inner red.

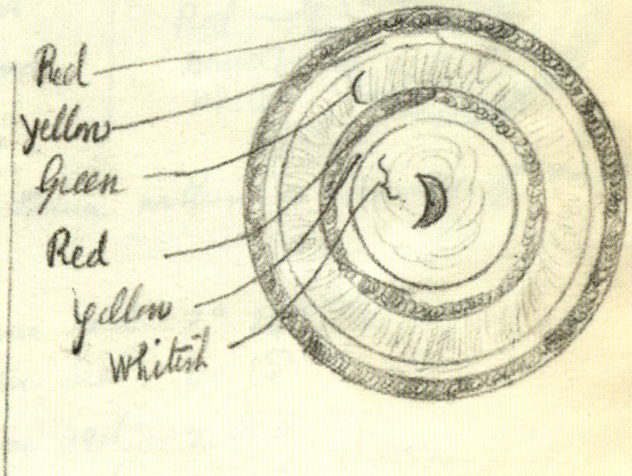

Yellow, both outer & inner, only a thin ~~str~~ strip; Reds & Greens broader and not so sharp.

Radius of inner red 2°. 29'
outer red 4. 43
Inner yellow 2. 4
Outer yellow 4. 8.

Page scanned from early Ben Nevis Observatory log book,
which show drawings of coronae with written descriptions, c.1884 Courtesy of Edinburgh Met Office / Marjory Roy

1884

very level as a whole on its upper surface
but broken into many small billowy ridges –
After sunset the sky just above this cloud
to NW was blood red, with yellow, green & blue above.
These colours continued all night following the sun
round from NW to NE & only disappeared at
Sunrise.

At 7 pm double or
triple glories seen
at East end of
hill top. Colours
as in diagram.
No 1. was very ~~faint~~
indistinct, a mere
blotch of colour. No 2 was the brightest of them.
No 3. was not so bright as No 2. and some-
-times was not seen; but when visible was quite
sharp & distinct.

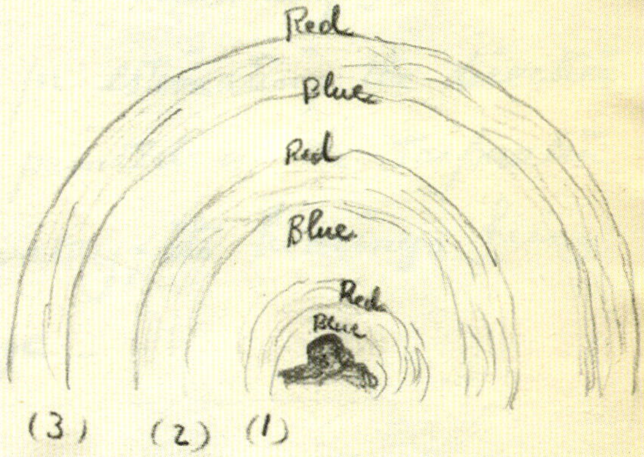

The following measurements were taken by Stefanome.

	Red (2)	Blue (2)	Red (3)
R.T.O.	3° 50′	3° 40′	6° 20′
A.R.	3° 50′	—	7° 0′

The glories were formed on the upper surface of the cloud
layer mentioned above.

Page scanned from early Ben Nevis Observatory log book, which show drawings of glories with written descriptions, c.1884 Courtesy of Edinburgh Met Office / Marjory Roy

WILSON'S RESEARCHES INTO CLOUD PHYSICS, LEADING TO DEVELOPMENT OF THE CLOUD CHAMBER

Inspired by his observations of glories and coronae on Ben Nevis, early in 1895 Wilson started work in the Cavendish Laboratory on experiments on clouds formed by the expansion of moist air, intending to study experimentally the optical phenomena which he had observed in the atmosphere. However, his experiments eventually took him into a very different field of research, that of atomic physics.

The amount of moisture which air can contain as an invisible gas depends on the temperature of the air – warm air can hold much more water vapour than cold air. If the air is cooled, saturation point is eventually reached and some of the moisture condenses to form water droplets. One method of cooling the air is to cause it to expand within a closed vessel.

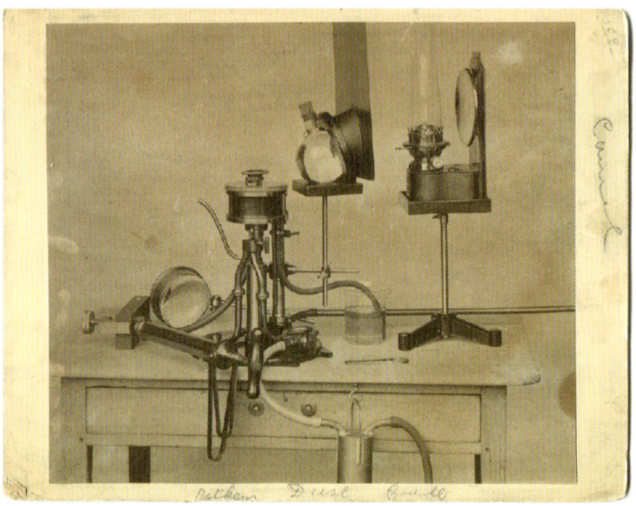

***John Aitken's dust counter instrument**, c.1890*
Courtesy of Edinburgh Met Office

Pioneering work by John Aitken of Falkirk showed that tiny cloud-sized droplets were normally formed with quite small expansions, but that if all the aerosols (solid and liquid particles, including dust particles) normally present in the atmosphere were removed, a very considerable degree of supersaturation was required for cloud droplets to be formed. Aitken devised an apparatus for counting the number of dust particles in a given sample of air; in this the cloud droplets formed by a slight expansion were allowed to fall on a grid and the numbers could be counted. A version of this apparatus was used at the Ben Nevis Observatory and although Wilson did not mention this in his later comments on his researches, he would have known of its existence.

Wilson started to work on air from which all of the aerosols had been removed by filtration through cotton wool. He discovered that under certain conditions, with larger expansions and greater cooling, cloudy condensation was sometimes observed. He found that ions (charged particles) could act as condensation nuclei and that the recently discovered X-rays gave enormous numbers of nuclei. In the absence of X-rays the condensation at sufficiently high supersaturations was due to the presence of ions in the air. It was not until 1911, however, that he devised a cloud chamber which could be used to photograph the tracks of ionising particles. Rutherford later remarked that this was 'the most original and wonderful instrument in scientific history'. For this achievement Wilson was awarded the Nobel Prize for physics in 1927.

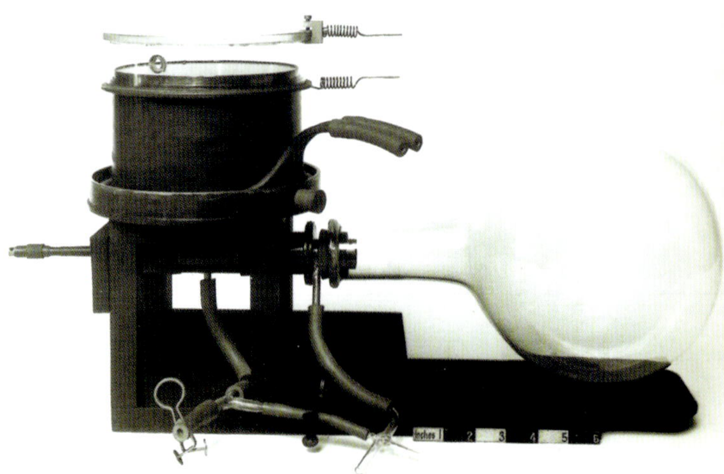

Wilson's first cloud chamber
Courtesy of The Science Museum, London

In his 1954 article he stated: 'The whole of my scientific work undoubtedly developed from the experiments I was led to make during my fortnight on Ben Nevis in September 1894. It is hardly necessary for me to say that these experiments might have had little result had it not been that they were made in the Cavendish Laboratory at the beginning of the wonderful years of the discovery of the electron, X-rays and radioactivity'.

RETURN TO HIS ROOTS

After his retirement Wilson moved to Edinburgh and then to Carlops, near the place of his birth, where he died in 1959. His retirement was an active one and he lunched weekly in Edinburgh with friends and colleagues. In 1973 a plaque was unveiled on the ruins of the Ben Nevis Observatory, to commemorate the achievements of the observers and the debt which Wilson owed to his short period of duty there.

The plaque reads:

"These are the remains of the METEOROLOGICAL OBSERVATORY which, with the path from Achintee, was built in 1883 with publicly subscribed funds. It was managed by the SCOTTISH METEOROLOGICAL SOCIETY and representatives of the ROYAL SOCIETIES OF LONDON & EDINBURGH. Hourly observations were maintained for 21 years and sent by telegraph via Fort William to meteorological services throughout Europe. The Brocken Spectre seen from here on cloud in the Great Corrie by CTR Wilson in 1894 led him to researches which culminated in the development of the Wilson Cloud Chamber, and for which he was awarded a Nobel Prize in 1927."

In 1996 the Institute of Physics, the Royal Meteorological Society and the Pentland Regional Park Authority erected a plaque to his memory at Flotterstone, close to Crosshouse Farm, where Wilson was born.

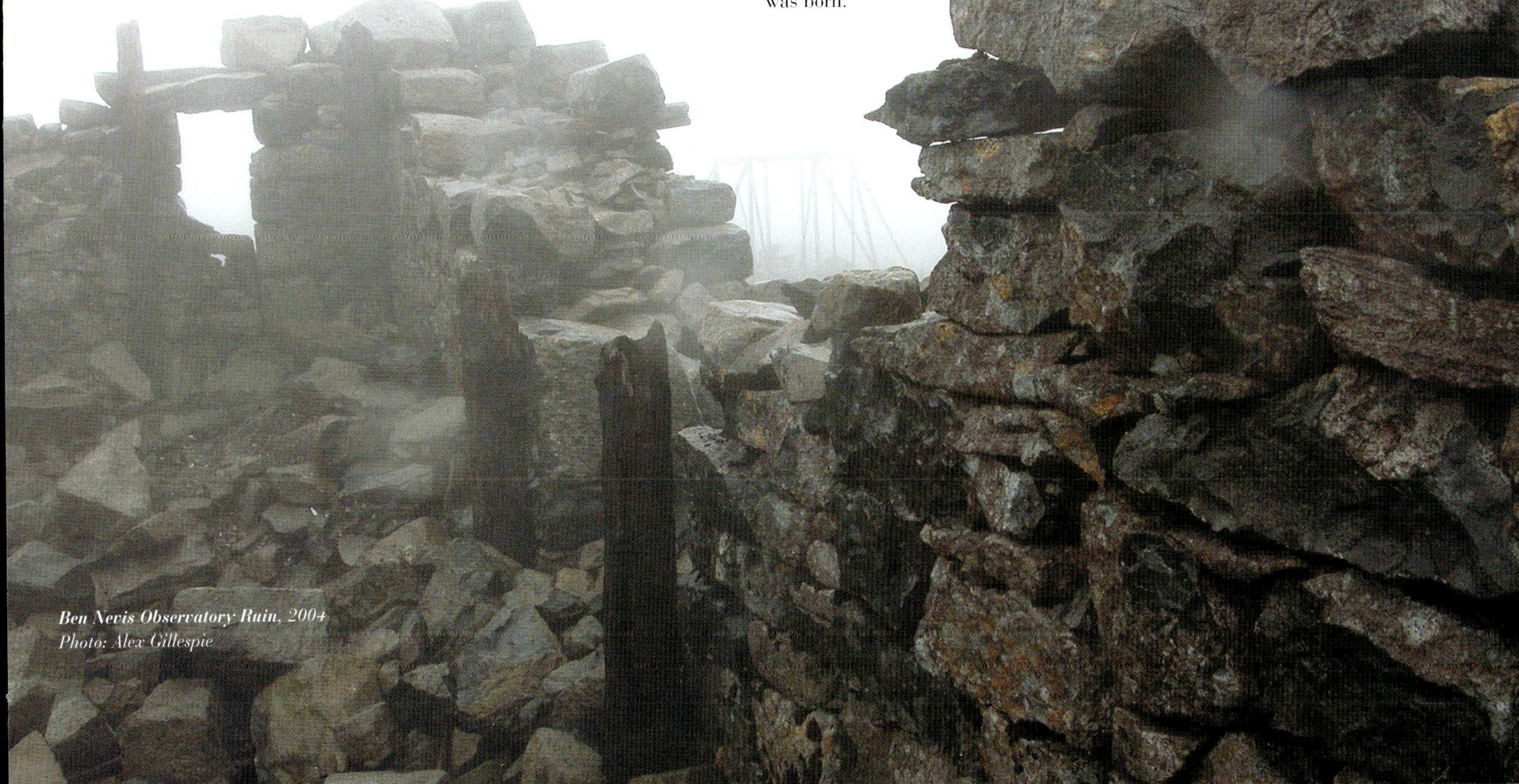

Ben Nevis Observatory Ruin, 2004
Photo: Alex Gillespie

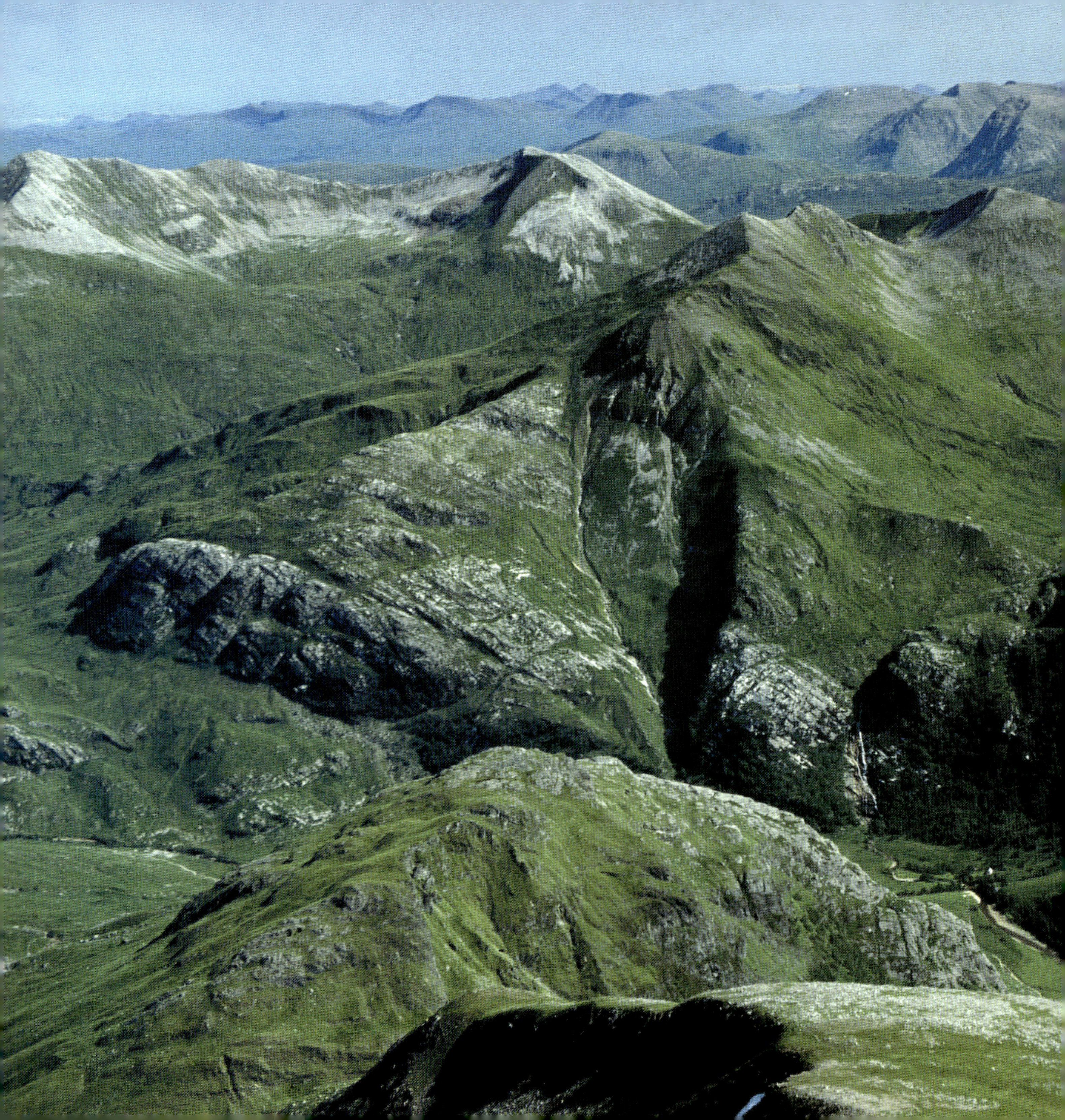

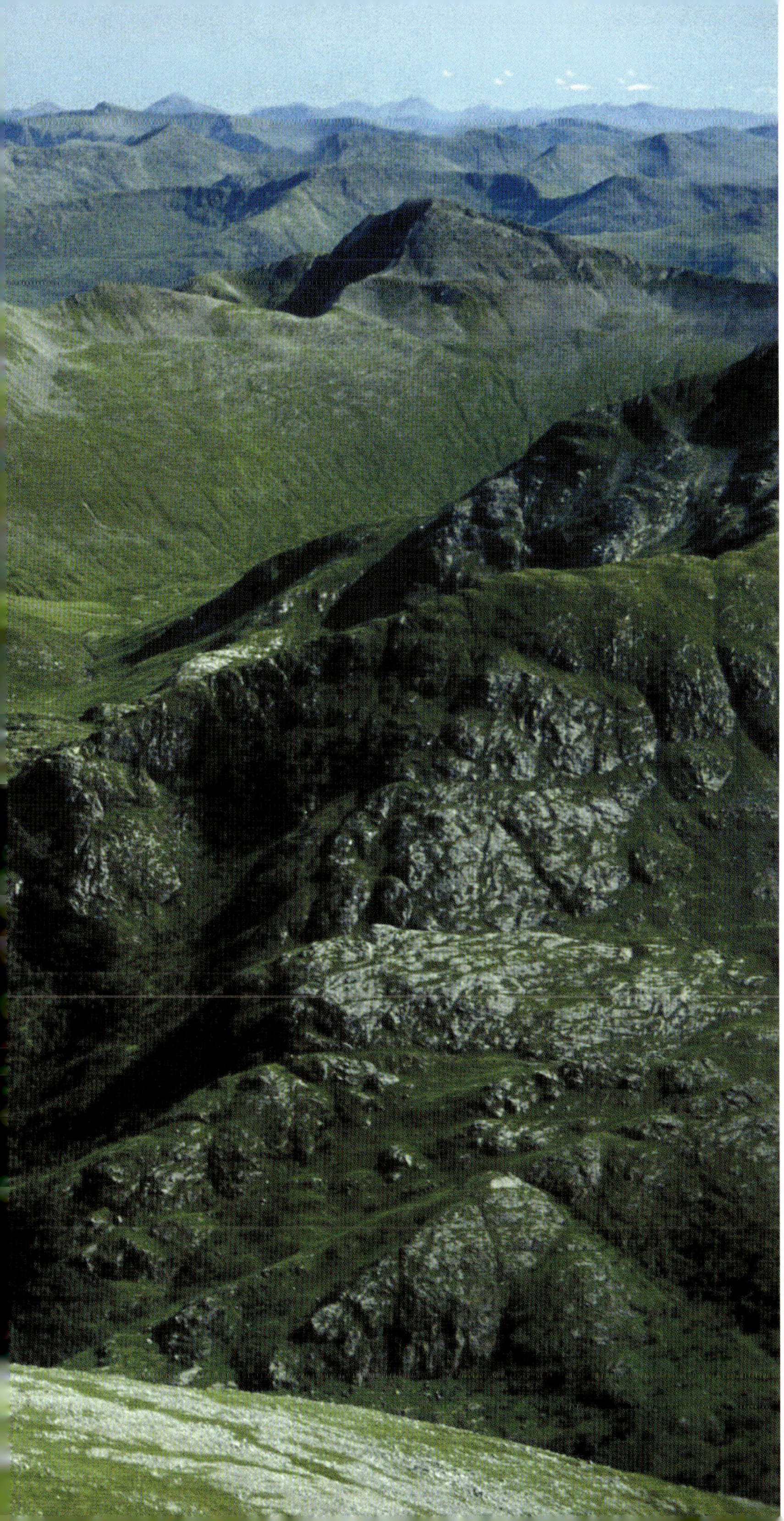

CTR Wilson's Legacy

A L MACKINNON

View from Ben Nevis summit looking south.
Photo: Alex Gillespie

October 2004 wasn't very different from any other October on the summit of Ben Nevis. Winds blew, rain fell, sun shone, the first snows of winter arrived and people went up and down Britain's highest mountain. Nothing unusual happened from the mountain's point of view. But that month was significant because it marked the centenary of the closure of the old Ben Nevis Observatory.

CTR Wilson is hardly a household name but as the only Scot ever to be awarded the Nobel Prize for Physics, and one who was stationed at the observatory for a period in 1894, he deserves a few minutes of our time. His Nobel-winning work stemmed directly from his time on Ben Nevis; many who might otherwise be unimpressed by his achievements will sympathise with his finding inspiration in such a place.

A family holiday at North High Corrie on Arran filled young Charles with awe at the beauty of nature, and he resolved to study the natural world in all its aspects. By the time he went to Cambridge in 1888, however, he had decided to study Physics, the science which tries to get to the bottom of all things.

His own recollections of his early years, published in the Notes and Records of the Royal Society are quite charming. 'I could not imagine what career I was fitting myself for as there were remarkably few openings for trained physicists. I felt I might be of use as an explorer as I had some knowledge of a wide range of sciences and powers of endurance tested on the Scottish hills. The prospects of gaining admission into an electrical engineering works seemed rather remote.' He was persuaded to settle down to hard work, and in 1892 completed his degree. A couple of rather unsettled years followed. A spell as a grammar school teacher was unsatisfactory, but his brief sojourn on Ben Nevis was to have more lasting consequences.

Young physicists were sometimes employed to relieve permanent Observatory staff for a week or two during the summer. The combination of scientific work and the remarkable setting must have appealed greatly to Wilson, and he duly spent two weeks in this role in September, 1894. This fortnight was to prove fateful – just as mountainous Arran had given him direction, so mountainous Lochaber led to the great achievements of his scientific career.

Anyone who has stood on a mountaintop and seen the Brocken Spectre knows what an intriguing phenomenon it is. The rainbow-like halo that forms around the head of the mountaineer's shadow has a slightly spooky beauty. On the summit of Ben Nevis in September 1894, CTR Wilson had often seen the Brocken Spectre and other glows and haloes, all results of sunlight passing through the tiny droplets of water in clouds. On his return to Cambridge he began experimental work aimed at understanding these elusive phenomena, constructing devices to produce little clouds in a controlled way in the laboratory. Working amongst people who were unravelling the nature of atoms, ions, X-rays and radioactivity, he was rapidly led in an unexpected and much more far-reaching direction. In due course, in 1911, the tracks of individual, subatomic particles were made visible as trails of water droplets in the Wilson cloud chamber. These particles had previously led only a phantom life in the minds of scientists, but in the cloud chamber they were rendered visible. The Nobel Prize in Physics, in 1927, was only the most renowned of the prizes and plaudits that followed this invention.

One evening in 1952, Donald Glaser enjoyed a glass of beer. Glaser, a physicist at the University of Michigan, had been using cloud chambers to study elementary particles. Gazing idly at the bubbles in his beer he asked himself why they formed at particular places in the beer, rather than all through it. He wondered if bubbles might form in the liquid at places where ions were present, in a way similar to water droplets in the cloud chamber. Then a chamber of such a liquid could function as a particle detector, revealing the tracks of subatomic particles as trails of bubbles. Being filled with a denser material than a cloud chamber it would reveal the tracks of much more energetic particles. Whether the beery myth of its origin is true or not, Glaser pretty rapidly produced a working prototype of the bubble chamber (not filled with beer! – bubbles in beer do not actually form at ions). In the years immediately following, bubble chambers took over from cloud chambers as the pre-eminent tool for the study of elementary particles.

Between its invention in 1911 and its retirement from research around 1960, the cloud chamber opened a window on the subatomic world. At first people saw things they had expected but even at first there were surprises. Eventually a world of perplexing complexity was revealed, and many Nobel Prizes were won for work carried out using Wilson cloud chambers. All of this may be viewed as Wilson's legacy, work that would have been difficult or impossible by any other means.

At the birth of the cloud chamber, Rutherford had just shown that the atom was almost entirely empty space. Positive electrical charge was concentrated in a central nucleus, and the much lighter, negatively charged electrons surround this nucleus, something like planets around the Sun.

Beta Rays
The cloud chamber picture shows many electron tracks, ejected from atoms as X-rays cross the chamber.
From Proceedings of the Royal Society A vol 104, p1. Published 1923

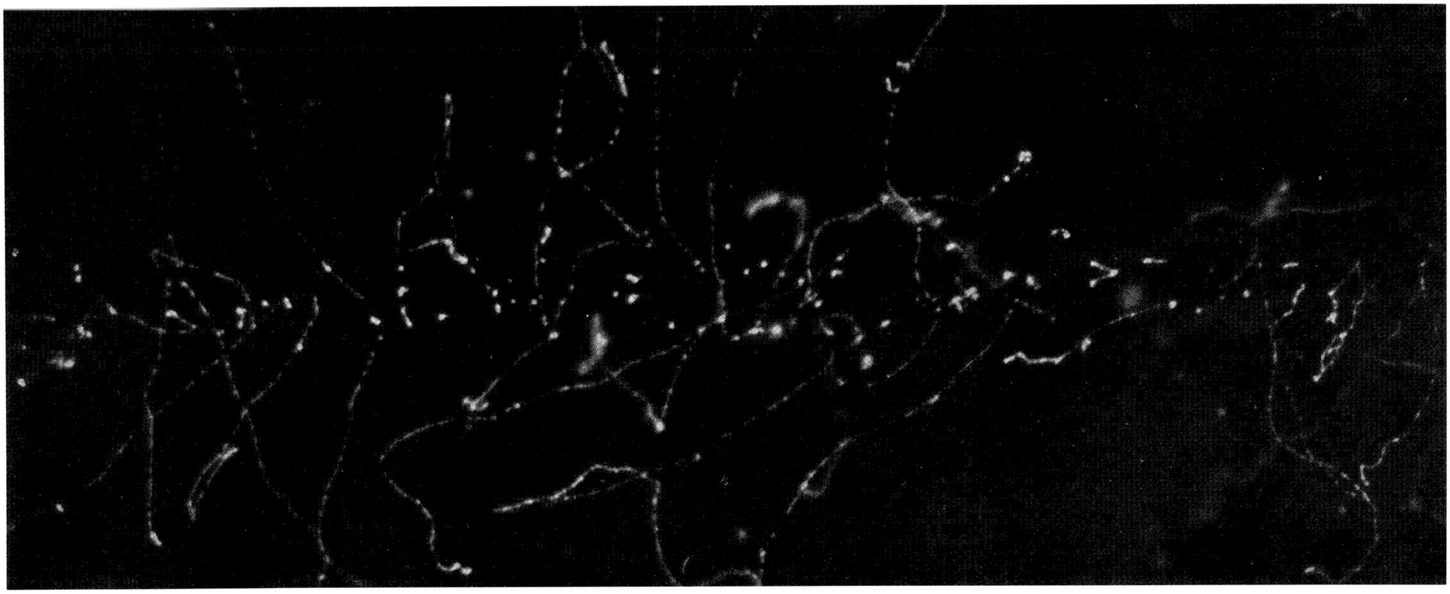

The subatomic pantheon grew rapidly. First, in 1932, the neutron was discovered. The nucleus of the atom is made up of neutrons, which have no electrical charge, and protons which are positively charged. The neutron leaves no cloud chamber track but the tracks of other particles recoiling from neutron impacts could be studied.

Most of what we know about subatomic particles has been learnt by making them collide more and more violently and inspecting the debris from these collisions. A comparison might be trying to deduce the workings of a Swiss watch by smashing it with a hammer and inspecting the cogs and wheels – difficult but not impossible. Cosmic rays (see below) provided a natural source of highly energetic particles, and study of their tracks revealed new sorts of particle. In 1932 Carl Anderson discovered the positively charged electron, the positron, in cloud chamber tracks. The positron, which we might also call the 'anti-electron', had been predicted theoretically by the Englishman Paul Dirac a couple of years earlier but Anderson was not aware of this. If a positron and electron come together they annihilate one another and we are left just with two particles of light, photons. But these energetic, penetrating photons can in turn spawn pairs of electrons and positrons and this remarkable phenomenon – light becoming matter, in the most vivid possible demonstration of Einstein's famous equation $E = mc^2$ – was also seen in cloud chambers.

Five years later Anderson, again studying cosmic rays in his cloud chamber, discovered another new particle. This so-called mu-meson turns out to be a sort of short-lived, massive version of the electron. About 1000 of these pass through your body every second. The small, happy family of the electron, proton and neutron began, slowly at first, to get bigger. In 1946 two Manchester scientists, Rochester and Butler spotted two tracks in the shape of a 'V' which must have been traveling away from the disintegration of an unknown particle. This new particle, now known as a K-meson, soon displayed surprising properties. The arrival on the scene of the K-meson marked the start of what has been called 'the baroque era of particle physics'. By the 1960s a 'particle zoo' had emerged, an apparent menagerie of particles with a whole range of masses and properties. These beasts were puzzling for some time, far too numerous and varied to be truly 'elementary'. In due course they came to be seen as manifesting the existence of a much smaller and simpler set of more elementary particles, the quarks.

Between his early work on ions and water droplets and the perfection of the cloud chamber in 1911, Wilson was kept busy with other tasks: teaching and demonstrating in the labs, and carrying out research on atmospheric electricity. All this time, in his own words, he 'used to think and dream of what the cloud method could be made to reveal. ..... They [the first cloud chamber tracks] revealed more than I had dreamed would be possible.' We see that this process continued for some decades, sometimes extending the known bounds of the possible.

New developments in science rarely arrive from nowhere. More usually some new development is in the air, everybody knows roughly what change is needed, and one or two people manage to reach that little bit further and make the necessary change real. Wilson's cloud chamber was an exception. Without the accidents of place and personality that come together in Wilson's story, it could have waited a long time for its invention. Not only his particular interests, but his vision, diligent persistence, and sheer skill in experiment were vital. Although the method was developed and speeded up by others, Wilson's early photographs were not surpassed for technical excellence. The cloud chamber and the discoveries made by others using it are certainly the most significant part of Wilson's legacy. He also made basic contributions to the understanding of thunderstorm electricity, however, and is credited with starting the study of radiation from the cosmos. In a sense, all of these interests come together under the heading of 'ions in air'. Intermingled at the time in Wilson's particular interests, these topics remain intermingled, often in surprising ways, to the present day. In these interests, in turn, we can see the lasting effect of his self-declared formative experiences, absorption in nature, time spent in the Scottish hills, on Ben Nevis in particular, but also on the Island of Arran at the age of 15.

Even in the absence of X-rays or other penetrating radiation, the few droplets produced in Wilson's cloud chamber testified to the continual production of a few ions in air at sea level. What agent could produce them? Wilson introduced a bold, imaginative speculation: Earth might be continually bombarded by some sort of immensely penetrating radiation from the wider cosmos. An experiment he carried out in a railway tunnel appeared to rule this idea out, but other people almost immediately pointed out that his experiment was flawed. While he himself played no further part in the direct study of cosmic radiation, he was the first to speculate that such a phenomenon might exist and the hare he set running was only chased down conclusively – in a balloon, at five kilometres altitude, with the reward of the Nobel Prize – by the Austrian Victor Hess. The study of cosmic rays is still an active topic in astrophysics, still posing fundamental puzzles. As we saw above, the study of the tracks of cosmic ray particles, revealed both in the cloud chamber and in photographic emulsions, led to several advances in the understanding of the subatomic world. In the summer of 1895 Wilson returned to Lochaber for a walking holiday. On the summit of Carn Mor Dearg, Ben Nevis' neighbour, he felt his hair stand on end and knew he should leave rapidly. He ran off the mountain to the floor of the corrie below, but this experience drew his attention 'very forcibly to the magnitude of the electric field of a thundercloud and to its sudden changes.' The resulting abiding interest, together with this skill and ingenuity as an experimenter, led to pioneering discoveries on thunderstorm electricity. He was the first person to reliably determine how much electric charge is transferred between cloud and ground in a lightning flash, constructing an ingenious, sensitive device to do this. He also deduced the distribution of electrical charge in a thundercloud: negative on its underside nearest the ground, positive higher up, as is now stated in all the elementary textbooks. In 1920 he made an important suggestion, that the electric field always present in the atmosphere might be maintained by thunderstorm activity.

It is almost surprising that something so familiar as the thunderstorm should remain incompletely understood. There is still debate over the mechanism by which electrical charge becomes separated in thunderclouds. Wilson thought long and hard about this question and contributed several basic ideas. These do not cumulatively explain thundercloud activity – although correct they do not operate rapidly enough – but they are still relevant to attempts to construct a successful theory, and they may explain the electric field underneath thunderclouds. In 1957, at the age of 87, he published his last article in the Proceedings of the Royal Society, the result of a lifetime's rumination on the mechanisms of thundercloud activity.

J J Thomson, the discoverer of the electron, was head of the Cavendish Laboratory in Cambridge when Wilson first worked there. He made some interesting remarks on the development of the cloud chamber, which also give us a little glimpse of Wilson's character:
"This work of CTR Wilson, proceeding without haste and without rest since 1895, has rarely been equalled as an example of ingenuity, insight, skill in manipulation, unfailing patience and dogged determination. Those who were not working at the Cavendish Laboratory during its progress can hardly realise the amount of work it entailed. For many years he did all the glass-blowing himself, and only those who have tried it know how exasperating glass-blowing can be, and how often when the apparatus is all but finished it breaks and the work has to be begun again. This never seemed to disconcert Wilson; he would take up a fresh piece of glass, perhaps say, 'Dear, dear', but never anything

stronger, and begin again...........It is to him that we owe the development of a method that has been of inestimable value to the progress of science."

Didn't scientists write beautifully in those days?

These words of JJ Thomson give us a glimpse of Wilson's character. Other glimpses are found in his own published reminiscences, which have great dignity as well as gentle humour, and also in his Royal Society obituary notice. There PMS Blackett, himself a Nobel Prize winner, wrote:
"Of the great scientists of this age he was perhaps the most gentle and serene, and the most indifferent to prestige and honour; his absorption in his work arose from his intense love of the natural world and from his delight in its beauties."

In a time when science is often seen as the enemy of nature, his life and work remind us that the natural world becomes no less beautiful because we understand it better.

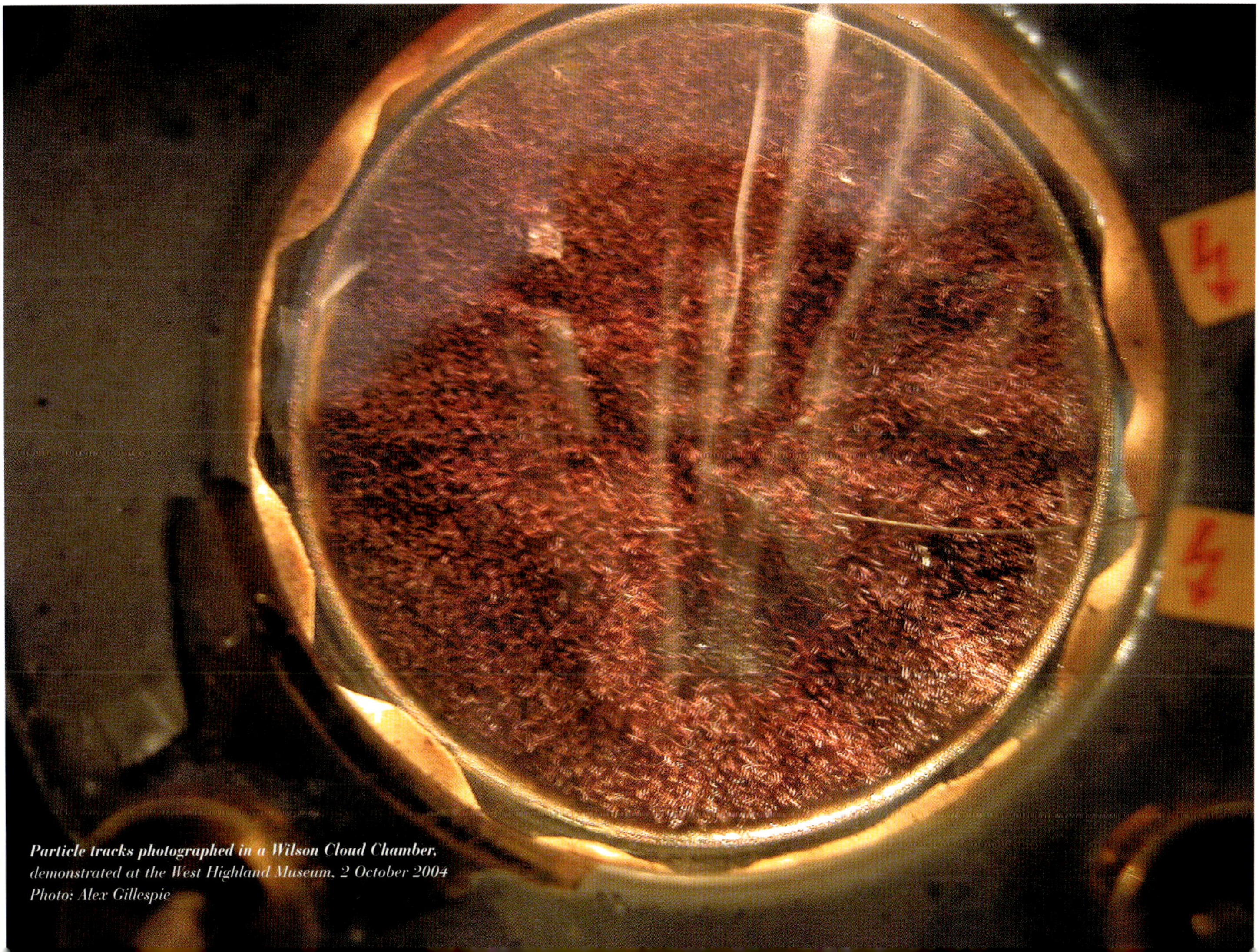

Particle tracks photographed in a Wilson Cloud Chamber,
demonstrated at the West Highland Museum, 2 October 2004
Photo: Alex Gillespie

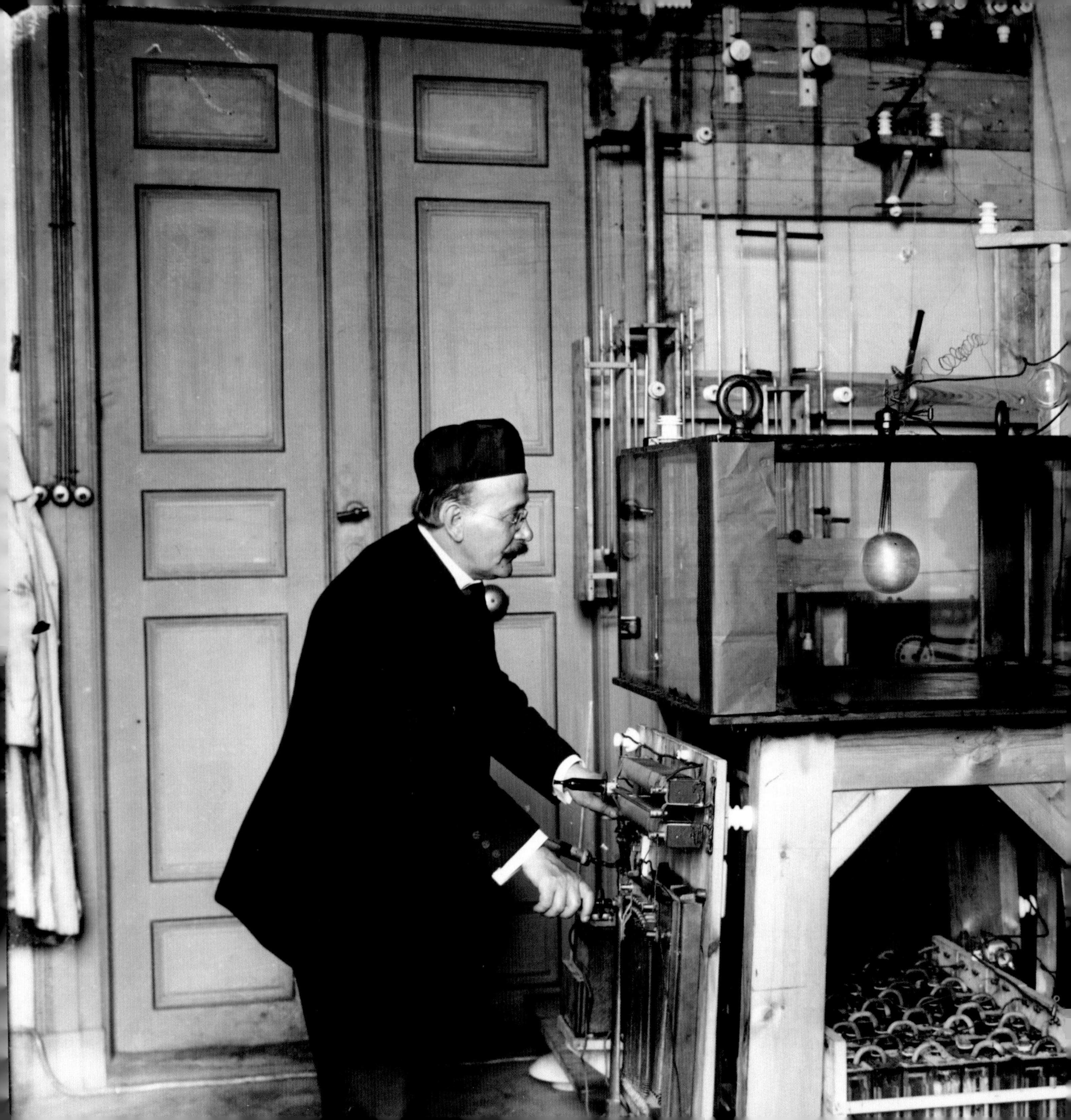

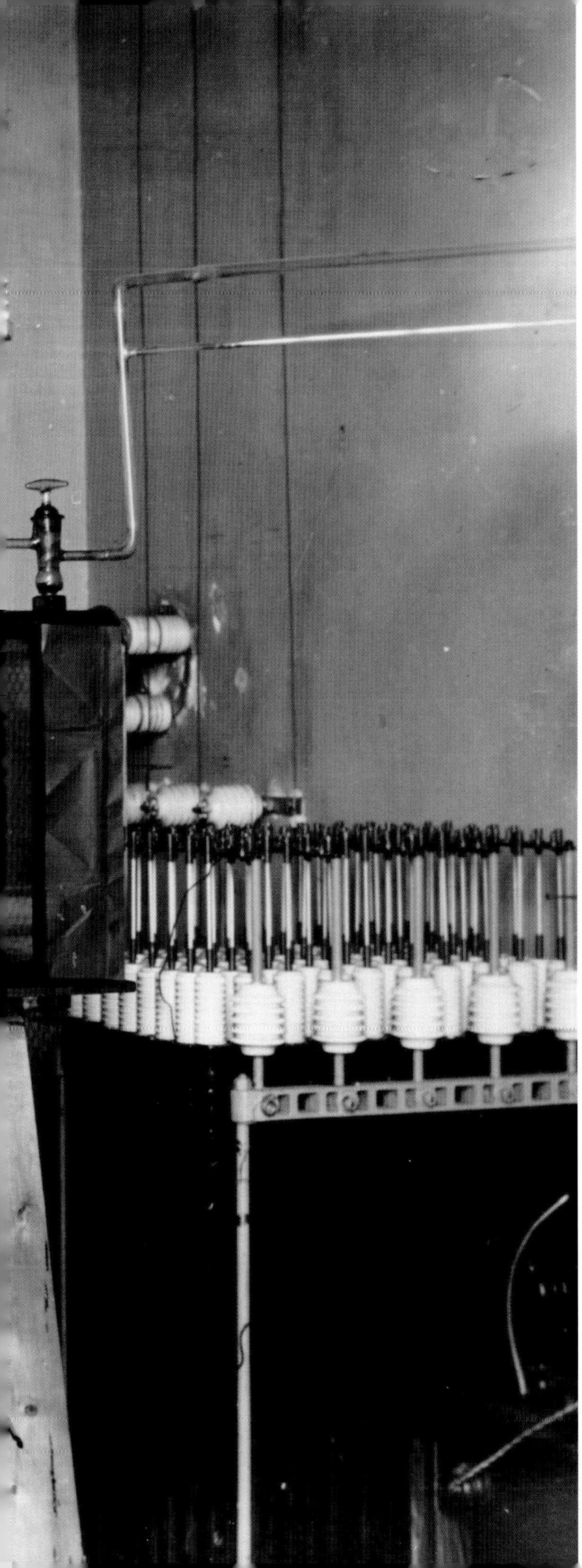

The Electric Universe: An Essay On The Scientific Life, Times, & Legacy Of Kristian Birkeland

STANLEY WH COWLEY
DEPARTMENT OF PHYSICS & ASTRONOMY, UNIVERSITY OF LEICESTER

Professor Kristian Birkeland in his laboratory at the University of Kristiania *(now Oslo) with his third 'prismatic' vacuum chamber of '320 litre' capacity, containing a spherical metal terrella. He is seen operating the variable resistance circuits that controlled the current passing through the wire coil inside the terrella, which generated the magnetic field surrounding it. The battery bank supplying the terrella current (up to 40 A) is seen underneath the wooden table supporting the chamber. The high voltage for the gas discharge inside the chamber was produced by a mains generator located in the basement under the laboratory, connected through the wiring seen on the wall above the chamber. To the right of the chamber is a capacitor bank of 80 'Leyden bottles', also used to generate discharges. The chamber was constructed in 1911 with help from his assistant Olaf Devik (1886-1987); the photograph thus dates from 1911 or 1912. The chamber and its base are now in the Museum of Science and Technology in Oslo.*
Photograph courtesy Nasjonalbiblioteket, Billedsamlingen, Oslo

Beginnings of Electromagnetism

We live today in an electromagnetic society. Anyone who doubts the truth of this statement should consider the consequences of turning off the electric power stations! No more instant source of energy to power homes, hospitals, and work places. Life would be much more labour-intensive, a return to every-day drudgery. Turning our back on electromagnetic waves would end our ability to communicate at speed across the world. No more texting and Channel 4! Electricity and electrical systems have played an increasingly dominant role in our lives over the past hundred years, and have been the decisive factor in creating the relative prosperity that now exists in developed societies. While these changes have also brought their problems, I doubt that proposals to return us to the pre-electromagnetic technical base of the mid-nineteenth century would find many takers after five minutes of realistic thought.

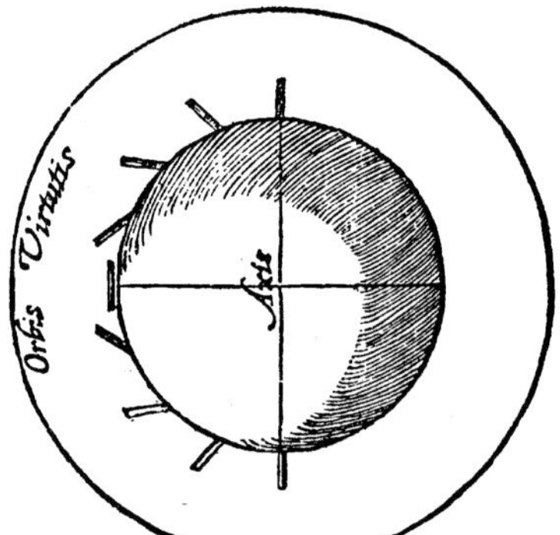

Fig.1(a)
Gilbert's diagram showing the direction of the magnetic field produced by a spherical loadstone at various points on its surface, as revealed by the directions taken up by small iron wires at various points. The field is parallel to the surface of the sphere on the loadstone's 'equator', and rotates to point in or out at the 'poles' lying on the magnetic 'axis'. The diagram is taken from Gilbert's treaties[1] of 1600, but has here been rotated on the page so that the magnetic axis points up and down.

The technical and engineering developments of the past hundred years were made possible by the physical and mathematical researches principally of the first half of the nineteenth century. Before that time electrical phenomena had mainly been known through the weak electrostatic effects that can be produced by rubbing certain insulating materials, such as amber and glass, while magnetism was known through the action of lodestones, naturally occurring lumps of magnetised iron oxide (magnetite). Lodestones were used by mariners to magnetise the soft iron compass needles used in navigation, through determination of the direction of the Earth's magnetic field. That the Earth itself is a giant magnet had first been proposed by William Gilbert (1544-1603), physician to Queen Elisabeth I, who published a treaties[1] on the topic in 1600. He likened the magnetic field of the Earth to that of a spherical lodestone, which he called a 'terrella' or 'little Earth'. Fig.1a, taken from his treaties, shows the direction of the magnetic field at various points over the surface of the terrella, as revealed by the directions taken up by small iron wires 'of the length of a barleycorn'. The field lies parallel to the surface of the terrella on the 'equator', but rotates to point radially in or out at the magnetic 'poles', lying on the 'magnetic axis'. The full field of a terrella in the space surrounding the sphere is indicated in Fig.1b. The magnetic field of the Earth is closely similar to this, being directed from the southern to the northern hemisphere, and with the magnetic axis being tilted by about 11° relative to the rotation axis, as suggested in the figure. Gilbert was the first to make a 'scientific' investigation of the magnetic effects of lodestones, and to clearly distinguish between magnetic and electrostatic phenomena.

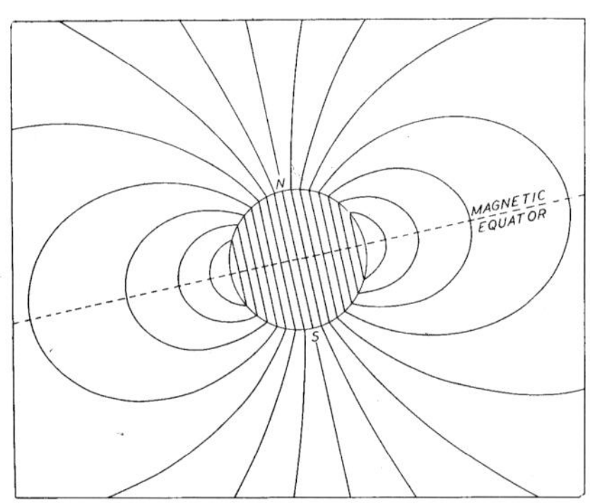

Fig.1(b)
Related diagram showing the extended magnetic field produced by a uniformly magnetised sphere (taken from Plate 1 in Ref.10). The Earth's field near the planet is of similar form, pointing generally from the southern (S) to the northern (N) hemisphere. However, the Earth's magnetic axis is tilted by about 11° relative to the rotation axis, as suggested by the tilt of the magnetic axis (and equator) in this diagram.

The first scientific connection between magnetic and electric phenomena was discovered by Hans Christian Ørsted (1777-1851) in 1819, who observed the magnetic effect of an electric current, a result rapidly followed up and extended by Jean Baptiste Biot (1774-1862) and Felix Savart (1791-1841), and by André-Marie Ampère (1775-1836). The most important discovery, however, was of the phenomenon of electromagnetic induction by Michael Faraday (1791-1867) in 1831. Induction is the process in which a varying magnetic field produces a circulating electric field, and hence a circulating electric current in the presence of an electric conductor. This phenomenon leads immediately to the principle of the dynamo, through which 'mains' electric power is generated to this day. Faraday made the first simple electric generator, and also the first simple electric motor, which is essentially a dynamo run in reverse.

That a phenomenon reciprocal to induction also occurs, in which a varying electric field produces a circulating magnetic field, was not realised until 1864. James Clerk Maxwell (1831-1879) proposed the existence of such an effect from theoretical reasoning, based on the proposition that electric charge is a conserved quantity. He also encapsulated the whole of known electromagnetic phenomena, including the new effect, into four mathematical equations, now known as Maxwell's equations[2]. This theoretical development was critical, because with the inclusion of the new effect, and only with its inclusion, it became clear that waves in the electromagnetic field could exist, which propagate in a vacuum, or other insulating medium such as glass or plastic, at very high speeds. The speed could be calculated from the theory, which for a vacuum turned out to be very nearly 300 million metres per second, equal to the observed speed of light within experimental uncertainty. Maxwell therefore proposed that visible light is an electromagnetic wave, the differing colours representing differing wavelengths (or frequencies) of the wave.

Measurement shows that the range of wavelengths of visible light is very narrow indeed, varying from 400 nm for violet light at one end of the spectrum, to about 700 nm for red light at the other (1 nm is one thousand millionth of a meter), while Maxwell's theory shows that in principle waves can exist of any wavelength at all. The first experiments to successfully generate and detect electromagnetic waves of much longer wavelength than visible light, that is to say radio waves, were performed by Heinrich Hertz (1857-1894) in 1888. This triggered the development of radio communications by Guglielmo Marconi (1874-1937) beginning in 1894, which culminated in the first trans-Atlantic transmissions in 1901. The fact that radio waves can span the Atlantic around the curve of the Earth was explained in 1902 independently by Arthur Kennelly (1861-1939) and Oliver Heaviside (1850-1925), who suggested the presence of an electrically conducting layer in the upper atmosphere. In the same year Oliver Lodge (1851-1940) suggested that this layer is produced by irradiation of the upper atmosphere by ultra violet light and X-rays from the Sun, which knocks electrons, temporarily, off atmospheric atoms. A population of positively charged ions and negatively charged electrons is thus created in the upper atmosphere – the gas becomes partially 'ionised' – and can carry an electric current if it is subject to an electric field. Extensive radio probing of the ionosphere, as it is now known, was begun by Edward Appleton (1892-1965) in 1924, who showed that it extends upward from a height of about 100 km.

Electromagnetic waves of much shorter wavelength than visible light, namely X-rays, were discovered by Wilhelm Roentgen (1845-1923) in 1895. These were produced when a high electric voltage was imposed between two metal plates (electrodes) in a glass tube from which nearly all the gas had been pumped to form a good vacuum. Under sufficiently high voltages the negatively charged electrode (the cathode) emits 'cathode rays', which produce a glowing 'discharge' as they interact with and ionise the residual gas in the tube. They also produce X-rays when they hit the positively charged electrode (the anode). In 1897 Joseph John Thomson (1856-1940) showed that these 'cathode rays' are negatively charged particles, much lighter than atoms, which became known as 'electrons'. These were the first 'sub-atomic' particles to be found, whose discovery, together with that of radioactivity, led eventually to an understanding of the structure of matter, and to nuclear and particle physics in the twentieth century.

The Scientific Work of Kristian Birkeland

The discoveries of Maxwell, Hertz, and Roentgen in the second half of the nineteenth century form the backdrop for the central topic of this essay, which is the scientific work and legacy of Kristian Birkeland[3] (1867-1917). Birkeland was born in Norway in 1867, a bright, ambitious, and obsessively hard-working person, whose first interests were in pure mathematics and theoretical physics. He graduated from the University of Kristiania (now Oslo) in 1890, and after a period spent teaching mathematics at a local high school, became full-time research assistant at the University in 1893. Shortly afterwards he was offered a scholarship, and used this to spend several months working as an assistant in leading universities in continental Europe, in particular with the mathematician Henri Poincaré (1854-1912) in Paris, and also with Heinrich Hertz and Philipp Lenard (1862-1947) in Bonn. However, the latter visit was cut short by the Hertz's death in 1894, after which Birkeland returned to Kristiania. His main research during this period was concerned with the use of Maxwell's equations to investigate the propagation of electromagnetic waves in various materials, and how energy is transported by the waves. In 1895 he obtained a teaching post at the University of Kristiania, and, stimulated by his continental visit, immediately began to set up an experimental laboratory to work on a new interest, the investigation of cathode ray discharges in gases[4], using high-voltage vacuum tubes similar to those employed by Hertz (following his more famous 'radio' experiments), Lenard, and Roentgen. Birkeland's special interest was the action of magnetic fields on such discharges, one of which he noted was the tendency for the cathode ray (electron) beam to be guided towards the pole of a magnet. In a leap of intuition, he used these observations to suggest a new theory of the 'northern lights' or aurora borealis, a phenomenon well known to people of northern origins, but whose cause remained mysterious and the subject of myth and superstition[5]. In a paper published in 1896, simultaneously in Norwegian, French, German, and English journals[6], he suggested that auroras are due to cathode rays emitted by sunspots on the Sun, which travel through the void between Sun and Earth, and are guided down to the polar regions by the Earth's magnetic field. When the rays hit the top of the atmosphere, he suggested that they make it glow like the glowing discharge produced in the residual gas in a vacuum tube (he did not know at what altitude), this being the aurora. Birkeland's 1896 paper marks the beginning of modern scientific thinking on the origins of the auroras, and a starting point of research on solar-planetary physics.

Birkeland's auroral theory may at first sound like a rather long and lucky shot. It was, however, supported by a number of known facts. First, it had been known since the work of Olof Peter Hiorter (1696-1750), his brother-in-law Anders Celcius (1701-1744), and their correspondent George Graham (1675-1751), in 1741, that the occurrence of auroras is connected with intervals of disturbance in the Earth's magnetic field, and that both occur simultaneously over wide regions of the Earth (in Uppsala and London in the first instance). This immediately suggests a large-scale electromagnetic phenomenon associated with electric current flow, the latter producing the magnetic effects. Second, the 11-year cycle of sunspot activity on the Sun had been recognised by Heinrich Schwabe (1789-1875) in 1843, while in 1852 a related periodicity had been found in geomagnetic disturbance by Edward Sabine (1788-1883), with magnetic disturbance increasing and decreasing with the numbers of sunspots. Since geomagnetic activity is linked with auroras, the occurrence of the latter also shows evidence of an 11-year cycle, and hence a link with sunspots. Third, while observing the Sun in September 1859, Richard Carrington (1826-1875) accidentally made the first observation of a large solar flare taking place within a sunspot group, a phenomenon that was followed 18 hours later by a large world-wide magnetic storm and auroral display on Earth. Thus a number of prior observations had been made linking solar activity and sunspots with geomagnetic disturbance and auroras, though without any clear idea about the causal agency of the link. Indeed, in 1892 the lack of a clear physical link led William Thomson (Lord Kelvin) (1824-1907), then President of the Royal Society in London, to dismiss its reality. Birkeland's 'cathode ray' theory suggested such a link, and simultaneously provided an explanation of why auroral phenomena are usually only found in annular zones surrounding the northern and southern polar regions (a fact firmly established for the northern hemisphere by Hermann Fritz (1830-1893) in 1881), via the focussing effect of cathode rays onto magnetic poles which he had observed in the laboratory. The theoretical explanation of the focussing effect was given in the same year as Birkeland's paper (1896) by Poincaré (with whom he maintained correspondence), who solved the idealised problem of the motion of individual electrons in the field of a single magnetic pole.

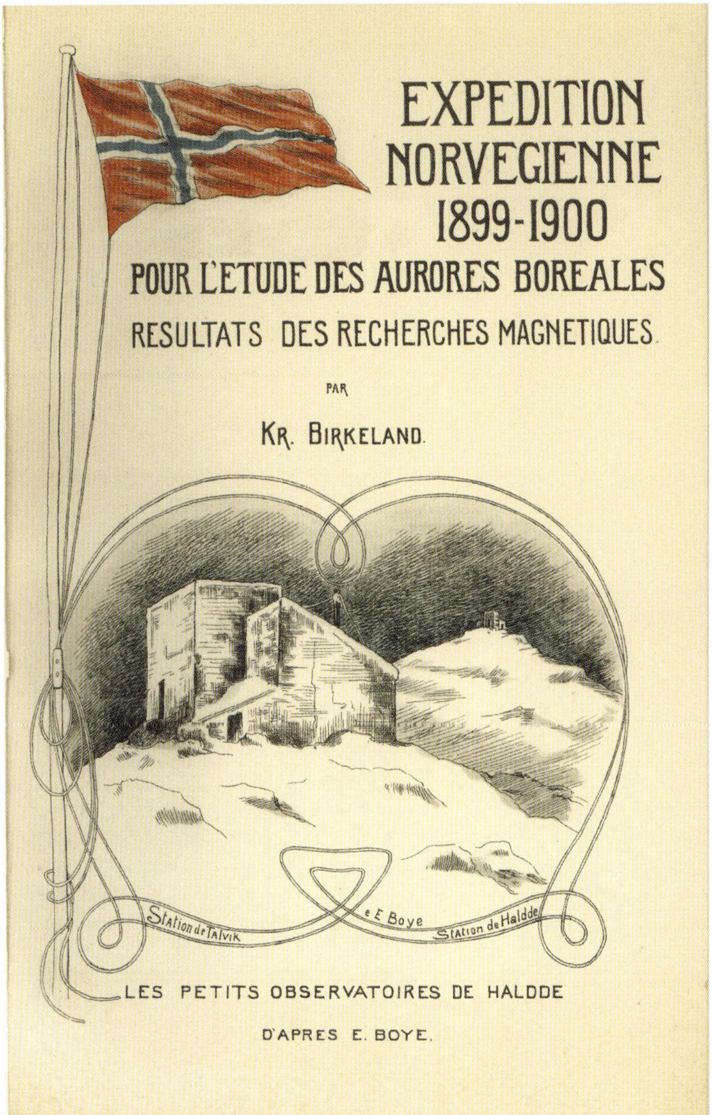

Fig.2
Title page of Birkeland's book[8] of 1901, describing the results of the 1899-1900 Norwegian polar expedition. The drawing shows the two mountain-top observatories at Talvik and Haldde, drawn by Birkeland's assistant Elisar Boye (1878-1900), who died in an avalanche at the end of the expedition.

In the years following his first auroral publication, Birkeland energetically followed up his idea in a number of directions. He first persuaded the Norwegian government to fund an expedition to make systematic observations of the auroras, geomagnetic effects, and meteorological conditions in northern Norway. This took place during the winter of 1899-1900. A particular purpose was to establish the heights of auroras by photographing them simultaneously against the star background from two observing points on the ground. For this reason two observatories were established on adjacent mountain tops, the main one at Haldde and the second at Talvik, separated by 3.4 km, but joined by a telephone line so that simultaneous observations could be organised. A drawing of these stations is shown in Fig.2, taken from the title page of Birkeland's volume[8] describing the results, published in 1901. These sites were chosen, despite their exposure to the elements in the arctic winter conditions, because it was maintained by some that auroras reached down to the mountain tops. On the other hand, the few attempts to deduce auroral heights that had been made previously[7] had found values above ~100 km (and thus in the ionosphere). In the event, the expedition showed that the auroras do not reach down to the mountain tops. Indeed, the 3.4 km baseline between Haldde and Talvik was not large enough to accurately deduce the heights, thus implying a significant altitude well above the realms of ordinary meteorology.

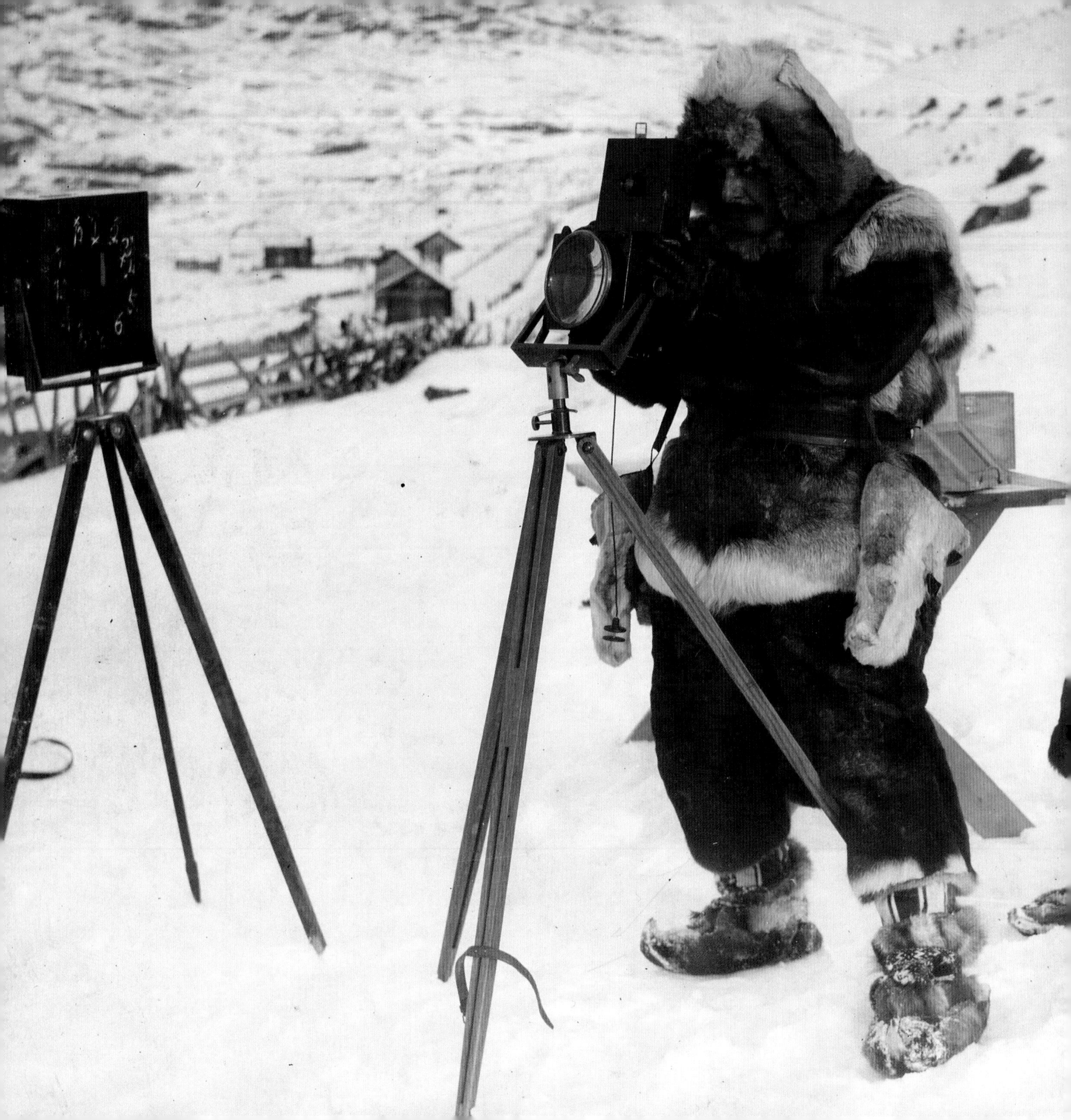

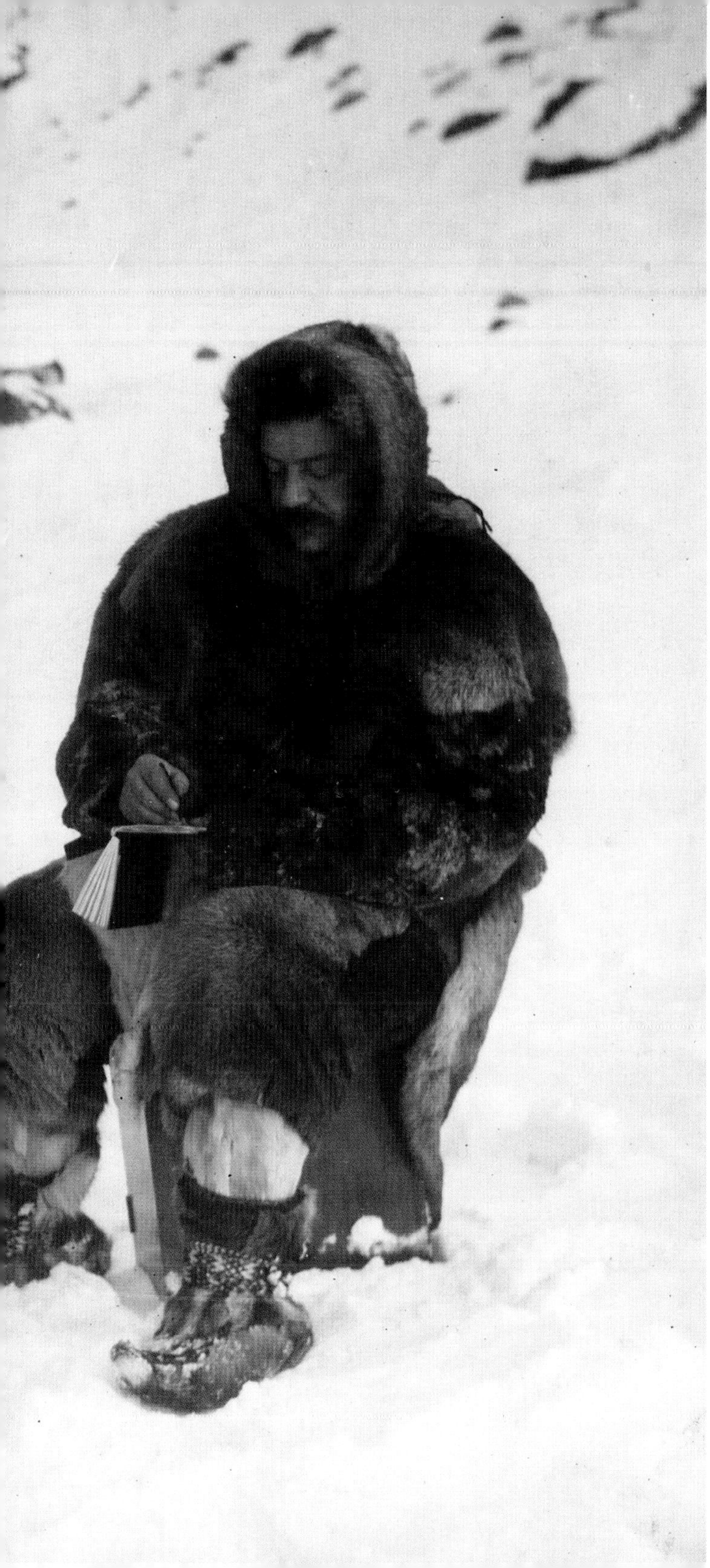

Further experiments in auroral photography were begun in 1909 by Birkeland's colleague Carl Störmer (1874-1957), leading to an auroral expedition to Bossekop[7] in northern Norway in 1910. Fig.3 shows Störmer with an auroral camera, together with his assistant Bernt Johannes Birkeland (no close relative) at Talvik, near Bossekop, on 25 February 1910. Rather amazingly, an album of the auroral photographs taken on this expedition still survives, one page of which is shown in Fig.4, taken on the night of 3-4 March 1910. This shows a series of images of quiet auroral arcs, together with other less distinct auroral forms. Successful systematic determinations of auroral height were eventually obtained during his second Bossekop expedition in 1913, using simultaneous photographic observations on a 27 km baseline. Full results were published in 1921, and showed that the auroras were confined to altitudes above about 100 km.

Fig.3
Carl Störmer with an auroral camera, together with his assistant Bernt Johannes Birkeland, during daytime preparations for night-time auroral photography at Talvik during the first Bossekop expedition[7]. The photograph was taken on 25 February 1910 by Anders Beer Wilse, a travelling professional photographer.
Photograph courtesy Norsk Folkemuseum, Oslo, negative number NFWB 02901

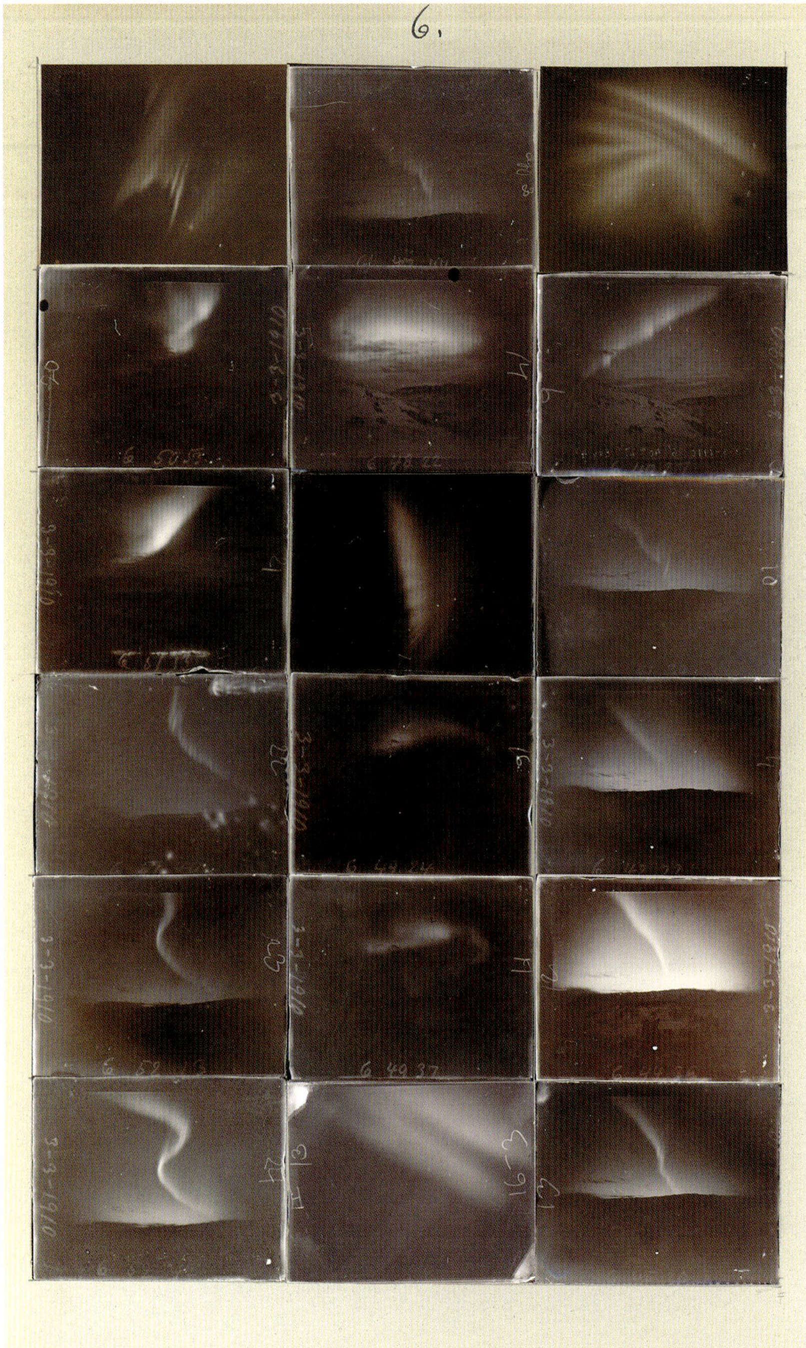

Fig.4
Page from an album of auroral photographs obtained during Störmer's first expedition to Bossekop[7] in northern Norway. It shows a sequence of images of quiet and rayed auroral arcs, together with other less distinct forms, obtained on the night of 3-4 March 1910. The camera (see Fig.3) employed an objective lens by Ernemann of Dresden, adopted from a small children's cine-camera, and sensitive 'Lumière etiquette violette' photographic plates[7]. With such equipment, the exposure time for bright auroral forms could be reduced to half a second.
Photo courtesy of Truls Lynne Hansen, Tromsø Auroral Observatory

A second consequence of Birkeland's sojourn on Haldde was his realisation that the hypothesised auroral electron (cathode ray) beams also formed an electric current, which could be the origin of the related magnetic disturbances. The current would flow down Earth's magnetic field lines into the upper atmosphere with the beam (or rather in the opposite direction because electrons are negatively charged particles), flow horizontally in the atmosphere in some manner for an interval (the existence of the conducting ionosphere was yet to be recognised), and then depart out of the atmosphere again to complete the electric circuit. In this way magnetic disturbance and auroral emissions would be inextricably linked, as observed. Because the currents and auroral displays are of large spatial scale in the polar regions, however, an extensive network of observations would be required to study them. Birkeland thus organised a second Norwegian polar expedition in 1902-03, with simultaneous magnetic and auroral measurements in Iceland, Svalbard, northern Norway (Kaafjord, near Bossekop), and Novaya Zemlya north of Russia in the Barents Sea (Fig.5). He also requested and received simultaneous magnetic records from magnetic observatories world-wide. The results, published in two major volumes[9] in 1908 and 1913, showed that bright auroral displays are associated with a systematic pattern of magnetic disturbance in the nightside hours. An example is shown in Fig.6, derived from data obtained on 15 December 1902. The arrows on the map indicate the direction and magnitude of the overhead electric current flow required to produce the magnetic disturbance observed on the ground. The pattern is consistent with an electric current flowing down the field lines in the east over Novaya Zemlya, then westward horizontally to the region beyond Iceland, and then up the field lines again. Birkeland had discovered a fundamental dynamical phenomenon in the terrestrial system which he called a 'polar elementary storm', now called a 'polar substorm'. He recognised these as being distinct from the positive and negative disturbances observed in more equatorial regions, which are known as world-wide 'geomagnetic storms', of which more later.

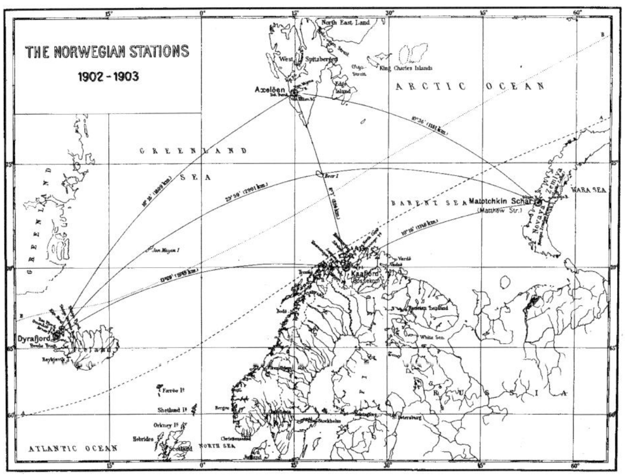

Fig.5
Map from the title page of Birkeland's book[9] of 1908 describing results of the 1902-03 Norwegian polar expedition, showing the location of the four magnetic and auroral observatories which he established. From west to east, these were located at Dyrafjord (Iceland), Axelöen (Svalbard), Kaafjord (near Bossekop, Norway), and Matotchkin Schar (Novaya Zemlya).

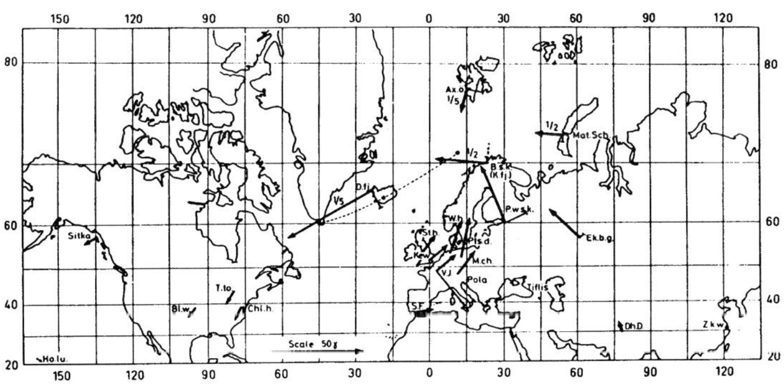

Fig.6
Map showing the magnetic disturbance observed on 15 December 1902 during the second Norwegian polar expedition of 1902-03, taken from Birkeland's volume[9] of 1908. Data from the four Norwegian polar stations can be seen in the upper part of the diagram (compare with Fig.5), to which have been added data from lower-latitude observatories obtained around the world, sent to Birkeland at his request. The magnetic disturbance is shown by the arrows which indicate the magnitude and direction of the overhead current required to produce the magnetic effect observed on the ground. The arrow at the bottom of the diagram shows the magnetic disturbance scale, corresponding to 50 nT (i. e. 50 'gamma' in pre-SI units), and we note that the disturbances observed at Birkeland's observatories have been divided by the fractions indicated.

It is worthwhile to mention here the cause of the significant delay between the end of the second polar expedition of 1902-03, and the publication of the first section of results[9] in 1908. This happened because Birkeland not only undertook basic research as described in this essay, but was also heavily involved in the practical exploitation of his electromagnetic knowledge. The most colourful of his inventions was an electromagnetic cannon which he never managed to persuade any of the armed forces of Europe to purchase and develop. The most useful was an electromagnetic furnace used to fix atmospheric nitrogen in the manufacture of calcium nitrate fertilizer, a development which led to the co-founding with Samuel Eyde (1866-1940) in 1905 of Norsk Hydo, Norway's first major industrial enterprise. This connection made Birkeland far richer and more famous in his day than his auroral researches ever did.

The second line of Birkeland's attack on the auroral problem was to expand his researches in the laboratory using cathode ray discharges in vacuum tubes. This work began immediately after the first expedition in the second half of 1900, and consisted in part of further investigations of the magnetic effects on gas discharges in relatively conventional vacuum tube set-ups, producing rayed band discharges that were likened to the aurora[4,8]. Much more significantly, however, he started a series of experiments that were completely novel, and would dominate his laboratory researches over the next 13 years. He created a cathode ray discharge between cathode and anode in a highly evacuated vacuum tube in the usual way, but then inserted a 'terrella' or 'little Earth' into the electron beam. The terrella consisted of a metal sphere, inside of which was an electromagnet consisting of an iron core around which a wire coil was wound, and through which an electric current was passed. The electromagnet produced a magnetic field outside the sphere that was designed to mimic the magnetic field of the Earth, like that shown in Fig.1. The term terrella itself was borrowed from William Gilbert, who had fashioned spherical loadstones to mimic the magnetic field of the Earth 300 years earlier[1], as mentioned above. Birkeland's terrella was painted with a phosphorescent paint which glowed in those regions where the electron beam reached the surface. When the current in the terrella coil was not turned on, such that it produced no magnetic field, the whole front surface of the sphere was found to glow where it was struck by the electron beam. When the current was turned on, however, and a substantial magnetic field generated (like that of Fig.1), Birkeland found that the glowing region became confined to two rings around each magnetic pole, showing that the beam was indeed deflected by the magnetic field and channelled into the polar regions as he had foreseen. These experiments were described in a chapter of his book[8] of 1901 whose main purpose was to present the results of the first polar expedition of 1899-1900. However, few details of the experimental arrangement were given, and no photographs either, since it appears that the vacuum tube accidentally broke before these could be taken.

These early experiments initiated a long series of related terrella investigations that lasted until 1913, the results of which were published in two instalments in the 1908 and 1913 volumes devoted mainly to the results of the 1902-03 polar expedition[9]. The results of the first series of experiments, published in the 1908 instalment, included an investigation into the seasonal variability of the 'auroral' rings, obtained by tilting the cylindrical vacuum chamber, and hence electron stream, while the terrella hung vertically at the end of a universal joint[4]. Experiments were also conducted in which the terrella was equipped with one or more screens, some with slots, holes, or attached pins, designed to determine the motion of electrons in the magnetic field. A set of eight photographs showing typical 'auroral emissions' excited on the surface of the terrella are shown in Fig.7. A parallel effort to calculate the electron motions theoretically was passed in 1903 to Störmer, who undertook the task of calculating the motion of charged particles in a dipole magnetic field (the field produced by two equal and opposite poles separated by a small distance, as in the field outside the sphere shown in Fig.1). Results were published by Störmer in a series of papers starting in 1904, but unlike the (unrealistic) problem of particle motion in the field of a single magnetic pole, solved by Poincaré in 1896, particle motion in the field of a dipole proved to be a much tougher problem, and could only be fully solved for individual trajectories by numerical integration of the equations of motion[7] (i.e. Newton's laws applied to charged particles).

After the second of the large (12 litre) glass vacuum tubes used in terrella experiments broke in 1908, Birkeland decided upon an improved experimental design, in which the vacuum vessel was of a rectangular 'fish tank' construction (which he called a 'prismatic chamber'). The top and bottom of the chamber was constructed from metal plates, while the sides were made of glass panes thick enough to withstand the outer atmospheric pressure. One reason for doing this was that the cylindrical glass tubes used hitherto tended to distort observations of what occurred within. Four such chambers were constructed between 1909 and 1913, with volumes increasing from 22 to 850 litres[4]. The chamber seen in the photograph on the title page of this essay is the third chamber of '320 litre' capacity (see Ref.4) constructed in 1911. In addition to such 'auroral' experiments, Birkeland became increasingly interested in other phenomena that could be produced using differing

Fig.7
Series of eight photographs taken from differing angles, showing terrella experiments which simulate the Earth's auroral rings surrounding the northern and southern magnetic poles. The phosphorescent paint on the surface of the metal sphere glows where it is struck by the electron beam emitted by the cathode in the vacuum chamber. The terrella suspension mechanism and electrical connections for the internal coil which produces the magnetic field can be seen at the top of some of the photographs. A set of longitudinal 'screens' can also be seen at the top of the terrella, which were used to help determine the electron paths before they hit the terrella surface.

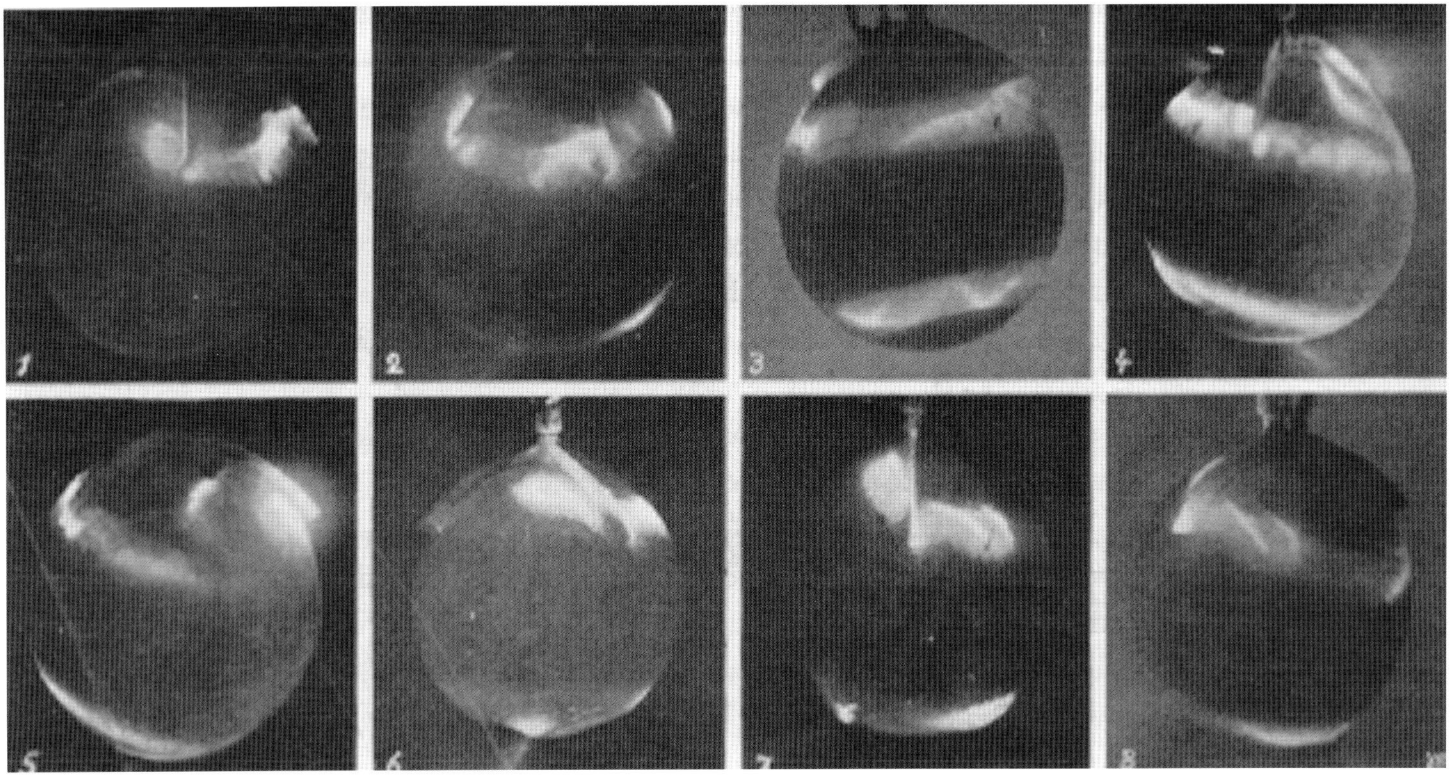

electrical configurations. He conducted extensive observations in which the metal surface of the terrella itself (without phosphorescent coating) was used as the electron-emitting cathode, thus, according to his ideas, modelling the Sun. With no terrella magnetic field he found that he could generate a roughly spherical discharge around the sphere, which he likened to the Sun's corona (the upper atmosphere above the visible surface of the Sun which can be observed during solar eclipses). When the surface of the sphere was roughened, the discharge from the surface became concentrated at numerous small points over the surface, which he likened to sunspots. When the magnetic field was switched on (by passing a current through the coil in the terrella), these hot-spots moved towards the magnetic equator, and eventually coalesced there to form either one or more equatorial rings, as shown in Fig.8, or a flat disc of light emission, depending on the parameters of the discharge. He suggested that the former represented a model of Saturn's rings, and the latter of the zodiacal light (a faint glow observed in the western sky after sunset and in the east before sunrise, mainly at tropical latitudes). On investigating the metal surface of the terrella after these experiments he found that it had become pitted, where material had been ejected from the surface by the discharge. He suggested that the planets and their moons were formed from material ejected similarly from their respective parent bodies. Thus from auroral beginnings, Birkeland's thesis eventually widened to encompass processes of cosmic significance. In 1913 he moved from Norway to Egypt to undertake observational studies of the zodiacal light, and died in 1917 in Japan, on his return journey. The last papers on which he was working before he died were lost in 1919, when the steamship Peking, which was carrying them between Japan and England, foundered in a storm.

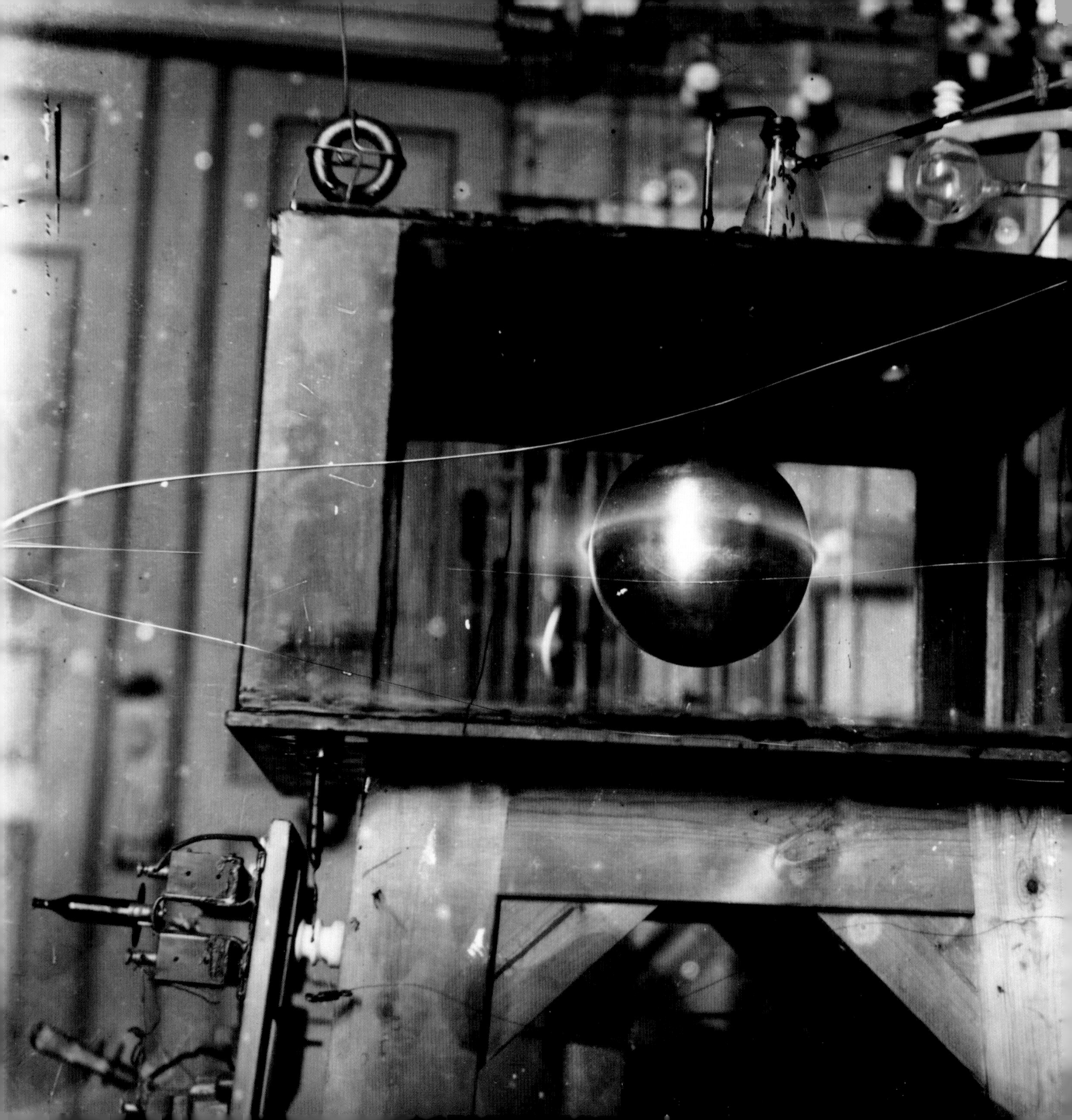

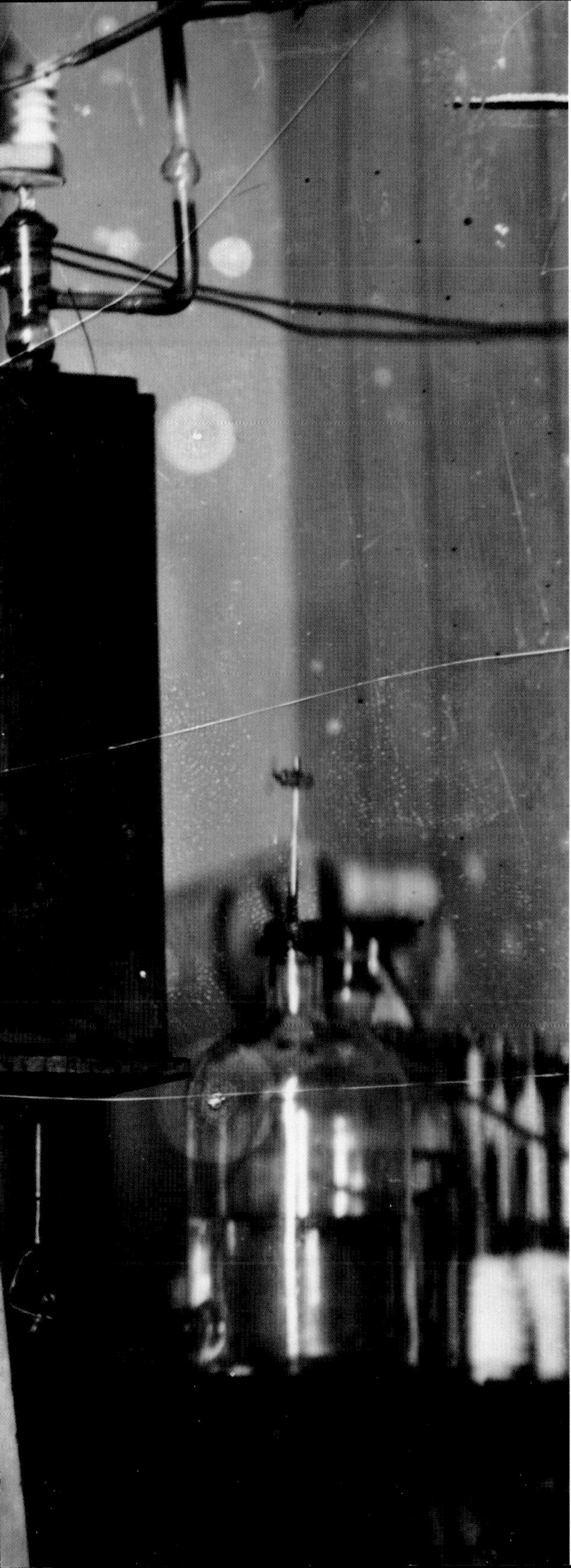

Most of Birkeland's detailed interpretations of the later terrella experiments are now known to be incorrect. Rings, moons, and planets are not formed by material constantly emitted from their parent bodies in cathode ray discharges, and neither is the zodiacal light. The latter, for example, is formed from a disc of dust lying in the ecliptic plane (the plane of the orbits of the planets), which is produced by decaying comets and illuminated by sunlight. It is nevertheless true that all the heavy elements of which we and our planet are composed were originally created by nuclear reactions in stars, flung out into space in supernova explosions to form the material out of which our solar system is composed. In addition, sunspots are not the seats of electrostatic discharges spewing forth material into the solar system, but are regions of relatively cool solar gas formed when ropes of strong magnetic field break through the Sun's visible surface. Nevertheless, the complex magnetic fields surrounding sunspot groups have been found to be the seat of massive gas eruptions from the solar corona into the solar system, called coronal mass ejections, which can cause geomagnetic storms. Though mistaken in its details, Birkeland's wider vision nevertheless provided the first glimpse of a cosmos in which not only gravity is important, but also electromagnetic phenomena.

Fig.8
Simulation of 'Saturn's rings' using a terrella of 24 cm diameter inside the third '320 litre' prismatic chamber (the actual internal capacity was 275 litres according to Ref.4). Photograph courtesy: Nasjonalbiblioteket, Billedsamlingen, Oslo

Aftermath

Following Birkeland's death in 1917, research on his specific idea that auroras and magnetic disturbances are caused directly by electron beams shot out from sunspots, guided down into narrow rings about the Earth's poles by the magnetic field, was carried forward by his colleague Carl Störmer, who continued to work on the problem of auroral heights and charged particle motion in a dipole magnetic field for much of the rest of his life. The results are collected in his book The Polar Aurora, written at the invitation of Appleton and published in 1955 when he was 81 years old[7]. In addition to Birkeland and Störmer's own work, the book also documents subsequent laboratory terrella experiments in Germany and Sweden, now largely forgotten. It also contains an excellent description of the 'succession of events in a common aurora borealis display', in fact that corresponding to a 'polar elementary storm' (polar substorm), in a chapter titled 'What the mathematical theory can or cannot explain' (Part II Ch. IX). The auroral dynamics is item six in the 'cannot explain' category.

The principal criticism of the theory, however, levelled already during Birkeland's lifetime by Arthur Schuster (1851-1934) in 1911, was that a beam of solar electrons of sufficient intensity to produce the magnetic effects observed would be rapidly dispersed under the mutual repulsion of the electrons' charge, well before it could reach the Earth. The relevance of calculating the trajectories of individual electrons was thus questionable. A resolution of this issue was first outlined by Frederick Lindemann (Lord Cherwell) (1886-1957) in 1919, who suggested that solar streams could instead consist of a gas composed of equal numbers of ions and electrons which is electrically neutral overall, and would not suffer from such disruption. Such a gas is called a 'plasma'. This idea was first taken up in the early 1930s by the mathematician Sydney Chapman (1888-1970) and his student Vincenzo Ferraro (1907-1974), and a little later by the physicist Hannes Alfvén (1908-1995), who between them contrived to generate apparently conflicting theories that led to thirty years of mutual disputation. The principal battleground was the theory of the origin of world-wide geomagnetic storms, these being relatively infrequent intervals, mentioned above, during which the Earth's magnetic field is disturbed over the whole globe, including the equatorial regions. Studies of world-wide magnetic records, such as those undertaken by Birkeland and afterwards by Chapman[10] and others, showed that geomagnetic storms generally begin with a sudden small increase in the Earth's field strength at equatorial and middle latitudes which lasts for a variable interval of a few hours, called the 'initial phase' (Birkeland's 'positive equatorial' perturbation), which is then followed by a more gradual but larger decrease which can last for several days before decaying slowly away (Birkeland's 'negative equatorial' perturbation). These world-wide magnetic effects are also usually accompanied by bright auroral displays and polar substorms at polar latitudes, but the aurora can also reach to much lower latitudes during such intervals than is usual.

Chapman and Ferraro considered the fundamental problem of what would happen when a large plasma cloud is emitted from the Sun and envelops the Earth. Their conclusion, published[11] in 1931, was that the Earth's magnetic field would not penetrate into the cloud, but would instead be swept up by the cloud and compressed on the dayside, as shown[10,11] in Fig.9a. This is a consequence of the high electrical conductivity of the cloud, made up of electrically charged particles. Fig.9a shows the Earth's magnetic field being compressed on the dayside as the solar plasma approaches from the right (compare with Fig.1), where the outer surface of the stream is taken initially to be a plane (the straight vertical line to the right of centre in the figure). Eventually, as the outer surface of the cloud sweeps past the Earth (at an estimated speed of ~1000 km s^{-1}), the magnetic field becomes confined to a cavity surrounding the planet, as shown in Fig.9b[10,11]. The size of the cavity was thought to be several Earth radii, determined by a balance between the pressure of the flowing solar plasma on one side of the boundary, and the magnetic pressure of the compressed planetary magnetic field on the other. This compression of the Earth's field was proposed as the cause of the initial 'positive' phase of the geomagnetic storm, which involves an increase in field strength in the equatorial regions. Magnetic studies suggested that the 'main phase', during which a much larger decrease in the field develops, is due to a ring of current flowing westward around the Earth within the cavity, in the equatorial region at distances of a few Earth radii, but the theory did not clearly indicate how such a current could be formed (see Fig.9b). Nor did it explain the related polar substorm currents and auroras. Chapman suggested[10] that some unforeseen link between Chapman-Ferraro and Birkeland-Störmer theory would be required to complete the picture.

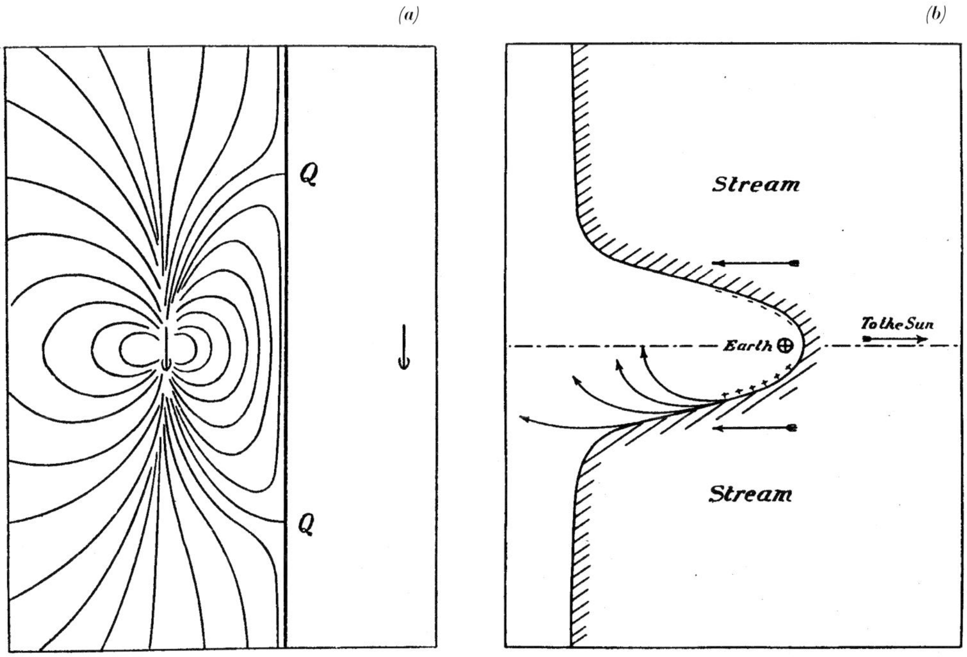

Fig.9
Illustration of the formation of a cavity in the plasma cloud or 'stream' flowing outward from the Sun by the Earth's magnetic field in Chapman and Ferraro's theory[10,11]*. Diagram (a) shows the compression of the Earth's magnetic field as the stream approaches from the right, where the outer surface of the stream is taken to be a plane, indicated by the straight vertical line in the cross-sectional plane of the diagram (the noon-midnight meridian plane). As this stream sweeps past the Earth, a cavity is formed in the stream surrounding the planet, as shown in diagram (b), which shows a cut through the equatorial plane of the Earth. The series of curved arrows on the left side of this diagram illustrates Chapman and Ferraro's suggestion for the formation of the main ('negative') phase of the storm, in which ions escape from the dawn-side surface of the stream into the interior of the cavity, carrying an electric current westward around the Earth. This is now known not to be correct.*

Alfvén's theory[12], published in 1939, also considered the effects of a stream of plasma flowing outward from the Sun, but envisaged that this would flow directly through the Earth's magnetic field without the formation of a cavity, associated with the presence of a large-scale electric field in interplanetary space, marked as *F* in Fig.10. His studies of the motion of charged particles in magnetic fields like the Earth's dipole field, however, showed that ions and electrons would drift on different paths near the planet.

The equatorial drift paths of negative electrons (electric force *-eF*) in his theory are shown in Fig.10a, where the direction of the Sun is at the top of the diagram. As the electrons approach the Earth, the magnetic field causes them to drift preferentially eastward via dusk, producing a westward current. The paths of positive ions are similar, but mirror reflected about the vertical axis through the Earth (the 'dipole' at the centre of the diagram), such that they drift preferentially westward via dawn. Both positive and negative particles thus contribute to an electric current flowing westward around the Earth. Alfvén suggested that this current, whose magnetic effect decreases the Earth's field in the equatorial regions, is the cause of the 'main phase' of the geomagnetic storm. He also suggested that the tendency of the ions and electrons to separate from each other on different paths would generate currents flowing along the Earth's magnetic field lines down to the ionosphere, which he thought were directed downwards by day and upwards by night, and were connected via horizontal currents flowing in the auroral ionosphere, this giving rise to polar magnetic disturbances. His suggested overall current system[12] is shown in Fig.10b. The auroras themselves were suggested to be due to 'a discharge carrying current from space to some conducting layer in the ionosphere'[7]. As we will see in the next section, the sense of the flows and currents in Fig.10 is now known to be incorrect (except for the eastward and westward drifts of electrons and ions, respectively, around the Earth). Nevertheless, Alfvén's theory was the first to suggest a connection between solar streams, equatorial and polar magnetic disturbances, and auroras, using relevant physical arguments.

These theories and their difficulties were elaborated and expanded upon by the principal protagonists and their colleagues from the 1940s to the early 1960s without clear-cut outcome. The principal fact that published discussions illustrate is that beyond a certain point, it was simply impossible to make decisive progress without direct observations of the space environment. Störmer's book[7] of 1955 concludes with the sentence 'The best way to get decisive information about the auroral corpuscles might be to send high-flying rockets directly up into the aurora, an experiment it might be possible to do now'. It is poignant that the era of direct measurements began with the launching of Sputnik 1 into orbit round the Earth in 1957, the year in which Störmer died. A revolution in understanding then took place over the next ten years, which swept away most of the previous controversies, as well, of course, as inspiring new ones.

Space observations showed first of all that the outflow of plasma from the Sun is continuous, and takes place over significant regions of the Sun's surface, though sudden outbursts can also occur from the regions above complex sunspot groups as mentioned above. The continuous nature of the outflow was first suggested in 1951 by Ludwig Biermann (1907-1986) from his study of the plasma tails of comets. The typical outflow speed of the plasma, termed the 'solar wind', is found to be about 500 km s^{-1}, such that it takes about three days to reach the Earth from the Sun. This flow is also found to carry the Sun's magnetic field out into the solar system, forming a weak but pervasive interplanetary magnetic field. Modern understanding of the supersonic solar wind outflow is founded in the theoretical work of Eugene Parker (1927-present), published in 1958. Space observations also showed that the Earth's magnetic field does indeed form a cavity within the solar wind plasma as envisaged by Chapman and Ferraro, now called the Earth's magnetosphere, and that since the solar wind flow is continuous, so is the existence of the cavity. However, the magnetosphere is found not to be empty, but to be populated by highly variable regions of flowing high-temperature plasma that respond strongly to interplanetary conditions, and which couples with the Earth's polar ionosphere to drive large-scale currents flowing between these regions as envisaged in outline by Alfvén. To understand how such a complex system comes about we need one final ingredient that was introduced by James Dungey (1923-present) in 1961, which forms the basis of modern solar-terrestrial research. The author was fortunate to have become Dungey's research student in 1968, and to have played a role in the development of these ideas since that time. We thus now examine what modern observations have revealed, and how they can be understood by combining the central ideas proposed by Birkeland and Störmer, Chapman and Ferraro, Alfvén, and Dungey.

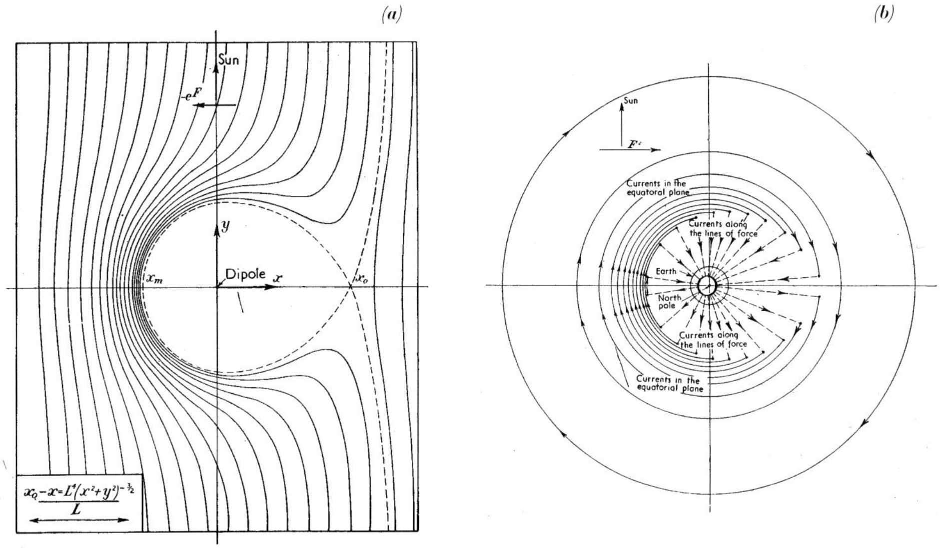

Fig.10

Diagrams illustrating Alfvén's theory of the main phase of a geomagnetic storm, taken from his paper[12] of 1939 (reproduced in Ref.7). Diagram (a) shows the drift paths of electrons in the equatorial plane, where the Earth is at the central point marked 'Dipole', and the electrons flow away from the Sun from the top to the bottom of the diagram. The fields considered are the dipole magnetic field of the Earth, and an electric field F directed from left to right (giving an oppositely-directed force on negative electrons of –eF as indicated). Diagram (b) shows the overall current system in Alfvén's model, consisting of westward currents carried by the drifting solar particles in the equatorial plane, field-aligned currents flowing downwards by day and upwards by night, and the closure current in the auroral ionosphere ring.

Solar-Terrestrial Physics

Observations from Earth-orbiting spacecraft show that the plasma structures formed by the flow of the solar wind past the Earth's magnetic field are as indicated in Fig.11. This diagram is comparable to Chapman and Ferraro's picture shown in Fig.9a, except that the Earth is now viewed from the other side, so that the Sun is to the left, and the solar wind flows from left to right. The black lines again indicate magnetic field lines, while the coloured dots represent various plasma populations. The green dots show the warm flowing solar wind plasma and populations derived from it, the blue dots the cool plasma from the Earth's ionosphere, while the red dots indicate a mixture of the two which is heated to very high temperatures (ten to a hundred million degrees) within the system. The Chapman-Ferraro boundary of the magnetosphere is indicated by the dashed line marked as the 'magnetopause', whose location is shown approximately to scale with the Earth. On this scale the Earth's atmosphere and ionosphere forms only a thin layer just above the surface. A 'bow shock' wave also forms in the solar wind ahead of the magnetopause boundary, because the magnetosphere represents a blunt obstacle in the supersonic solar wind flow. One innovation in this picture compared with Chapman and Ferraro's, however, is that the solar wind is shown as also having a magnetic field threading through it. To a first approximation plasmas and magnetic fields behave as though they are 'frozen together', a fundamental concept in space plasma physics that was first clearly enunciated by Alfvén in 1942, as a consequence of Faraday's law applied to a highly electrically conducting plasma. (Alfvén used the concept in the discovery of a new type of wave, now called an 'Alfvén wave', which transfers plasma motion from one place to another along magnetic field lines.) Consequently, the solar wind carries the Sun's magnetic field out into the solar system, as mentioned above, to form a weak but pervasive interplanetary magnetic field as shown. This has been depicted in the diagram as pointing to the south, that is, downwards. Observations show that the field is variable in direction, and changes on times scales from a few minutes to a few hours and days, so this represents one possible orientation that is present part of the time.

The existence of the magnetospheric cavity containing the Earth's field, bounded by the magnetopause, is a consequence of the validity of Alfvén's 'frozen-in' theorem as a first approximation. The solar wind and interplanetary magnetic field are frozen together, as are the Earth's magnetic field and magnetospheric plasma, such that when the two plasmas meet they do not mix, but form a thin boundary between them (the magnetopause) whose position is determined by pressure balance. The existence of such plasma 'cells', separated by thin boundaries, is ubiquitous in cosmic plasmas. However, while the frozen-in theorem is thus a powerful concept in large-scale plasma systems, it represents an approximation that can break down when plasma conditions vary rapidly in space or time, a limitation Alfvén himself never tired of pointing out in his later years. The magnetopause boundary represents one such region, where conditions vary rapidly between the post-shock solar wind plasma and interplanetary field on one side, and the magnetospheric field and plasma on the other. In 1961 James Dungey suggested that the break-down of the frozen-in condition at the boundary would allow the solar wind and planetary magnetic field lines to connect to one another across the boundary, as shown on the left-hand side of Fig.11. This process of 'magnetic reconnection' as Dungey called it[13], forms 'open' field lines which connect across the boundary from the solar wind at one end, to the Earth's polar ionosphere at the other. The solar wind 'ends' are carried away from the Earth by the plasma flow, and are stretched out on the night side to form a comet-like magnetic 'tail' that extends typically about a thousand planet radii away from the Earth (and thus far beyond the end of the diagram in the figure). Periodically, however, typically every several hours, open field lines undergo reconnection again in the centre plane of the tail. This forms new closed field lines on the earthward side of the reconnection region which collapse back towards the Earth, while 'disconnected' field lines flow back into the solar wind on the other, as shown on the right side of Fig.11. Plasma on the collapsing closed field lines becomes strongly compressed and heated, forming the hot plasma populations shown by the red dots in the figure. Such episodes of rapid large-scale tail collapse are called 'magnetospheric substorms', and produce the 'polar substorms' or 'polar elementary storms' first by recognised by Birkeland.

Overall, the transport of open field lines by the solar wind from the dayside to the tail, followed by the episodic return flow of closed field lines from the tail back towards the dayside, forms a cyclic flow of plasma and field inside the Earth's magnetosphere which is called the 'Dungey cycle'[14]. This flow is not steady, however, but is strongly modulated by the direction and strength of the interplanetary magnetic field. When the interplanetary field is strong and points to the south opposite to the Earth's equatorial field as shown in Fig.11, rapid reconnection occurs at the dayside magnetosphere, and strong Dungey-cycle flow is excited. However, when the interplanetary field is weak or points to the north then dayside reconnection is much reduced, and with it the Dungey cycle flow.

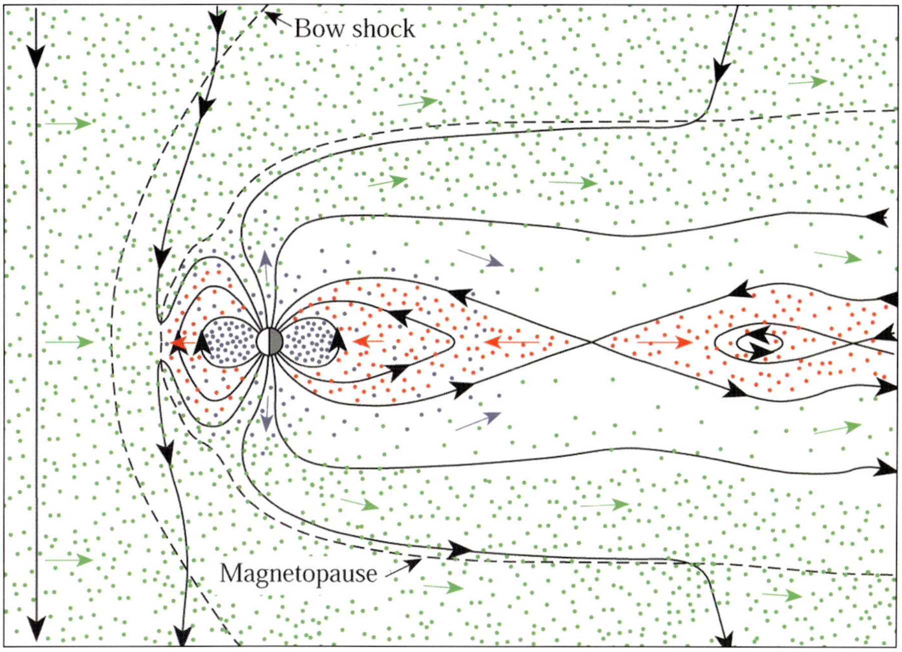

Fig.11

Diagram showing the principal features of the Earth's plasma environment as observed by Earth-orbiting spacecraft, in a cut through the noon-midnight meridian plane, with the direction toward the Sun to the left. The black solid lines with arrows show magnetic field lines corresponding to the Dungey model, while the dashed lines show the bow shock and magnetopause as marked. The coloured dots show the principal plasma populations originating in the solar wind (green) and the Earth's ionosphere (blue). Both sources contribute to the hot 'plasma sheet' population located on outer closed field lines (red).
From Cowley et al. , Phil. Trans. A, 361, 113, 2003

The key feature of the Earth's plasma system is therefore the strong modulation of its behaviour by the direction of the interplanetary magnetic field. The complicated plasma phenomena that occur in the magnetopause boundary and in the tail, associated with the Dungey cycle, are the prime focus of current space programmes such as the Cluster mission of the European Space Agency (ESA). Four Cluster spacecraft are presently flying in formation in orbit around the Earth (Fig.12a), passing repeatedly through these plasma structures every two and a half days. These were launched two at a time on Russian Soyuz-Fregat boosters in July and August 2000 (Fig.12b), and were assembled into a pyramid-shaped formation once in orbit. Each spacecraft carries an identical payload of scientific instruments to measure the magnetic and electric fields in space, together with the plasma particles. Independent measurements from the four spacecraft allow us to determine the orientation of the plasma boundaries, for example, and how they move across the spacecraft. Four separated magnetic field measurements also allow us to estimate the direction and strength of the electric currents flowing in space, carried by the plasma particles.

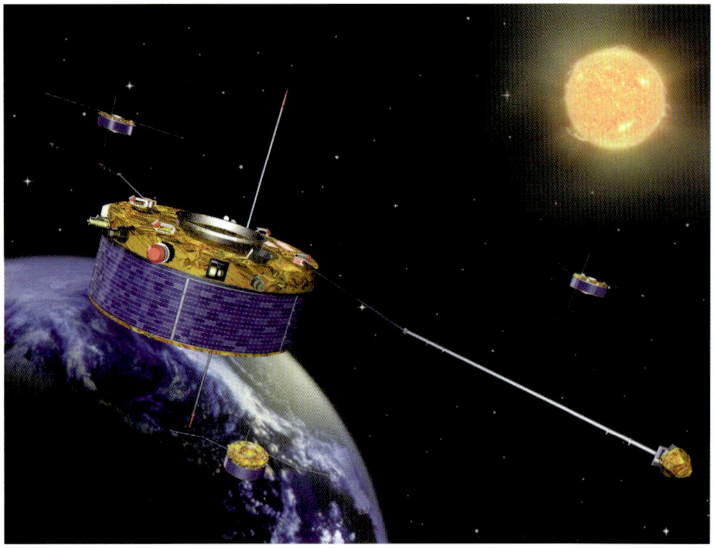

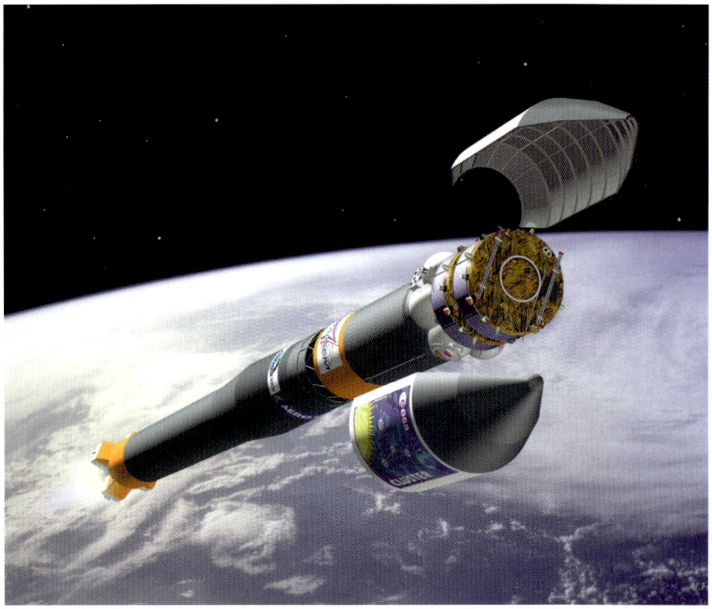

Fig.12(a)above
Impression of the four Cluster spacecraft flying in formation around the Earth, observing the interactions that take place between the out-flowing solar wind and the Earth's magnetic field.

Fig.12(b)below
Impression of the moment during launch when the nose fairing of the Soyuz booster was released to reveal two Cluster spacecraft and the Fregat fourth stage on the top of the rocket stack.

Images courtesy of the European Space Agency

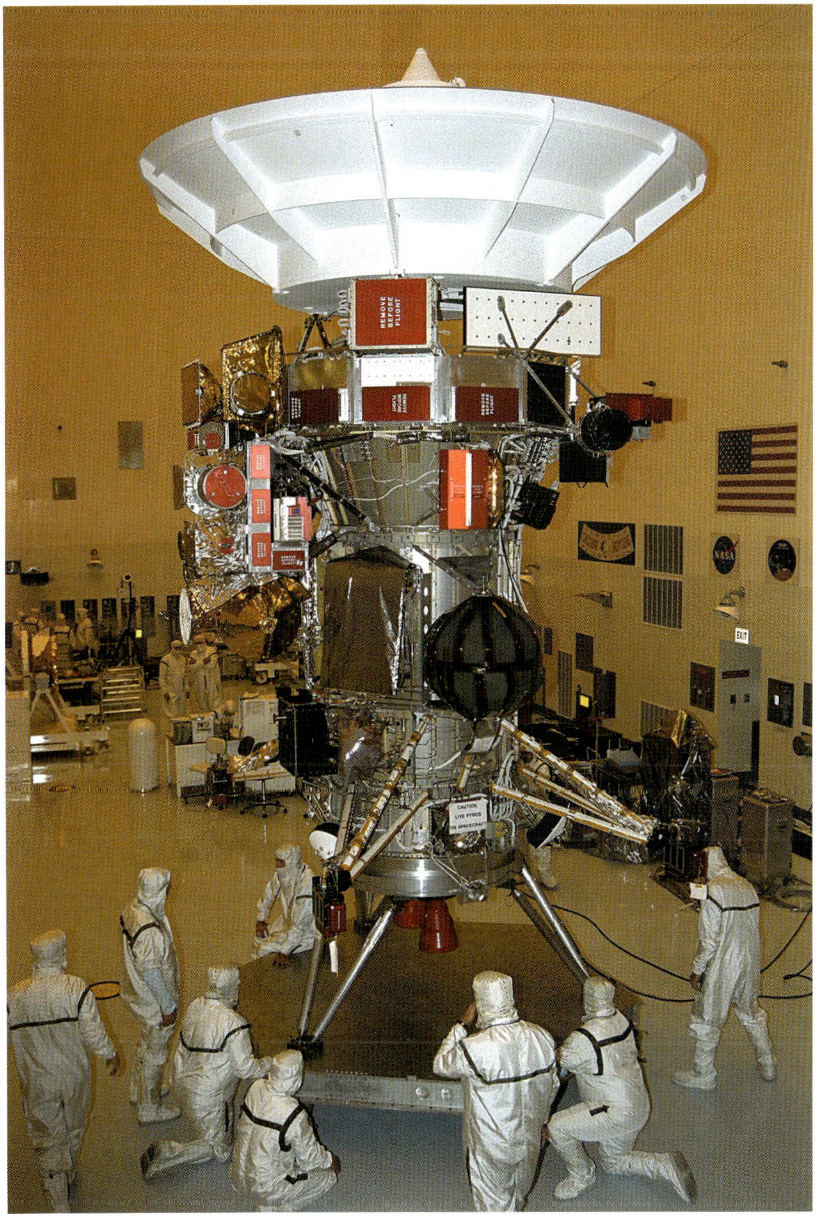

Explorations of the environments of other planets in the solar system are also in progress, one of which is the NASA-ESA Cassini-Huygens mission to Saturn and its moon Titan. Fig.13 shows the Cassini spacecraft being assembled at the Kennedy Space Center prior to launch in October 1997. After sling-shot fly-bys of Venus, Earth, and Jupiter, the spacecraft eventually arrived at Saturn in July 2004, and fired its rocket motor to place it into orbit around the planet. Several years of observations of Saturn's magnetic field and plasma populations with the Cassini orbiter are now in prospect, together with a landing on Titan by the Huygens probe in January 2005.

Fig.13
Technicians from the Jet Propulsion Laboratory reposition and level the Cassini orbiter in the Payload Hazardous Servicing Facility at the Kennedy Space Center in July 1997. The white structure on top of the spacecraft is the high-gain antenna used for radio communications, under which is the equipment module which carries the scientific instruments. These sit on top of the propulsion module, which is a rocket motor used to slow the spacecraft to put it into orbit around Saturn in July 2004.
Photo courtesy of NASA.

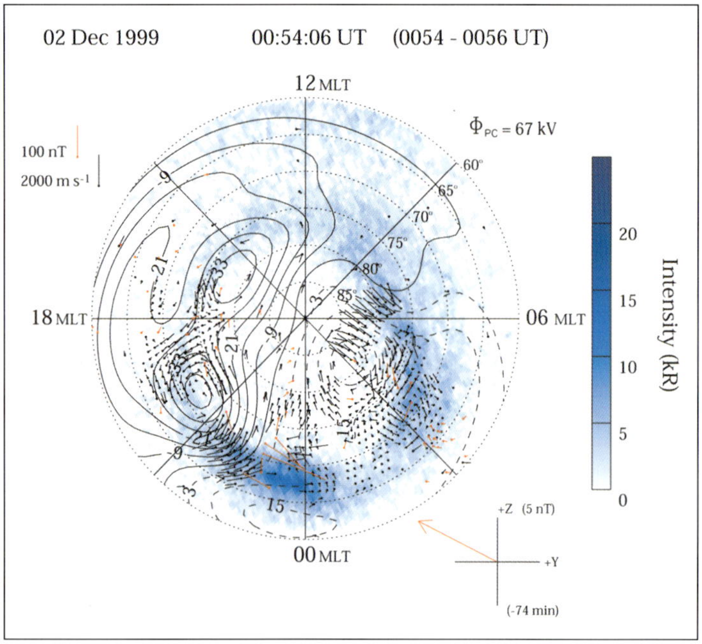

Fig.14
View looking down onto the Earth's northern polar region showing auroral emissions observed in ultraviolet light by the VIS camera on the Polar spacecraft (on a 'blue colour' intensity scale shown on the right). The magnetic pole is at the centre of the diagram, noon (the direction towards the Sun) is at the top, dusk to the left, midnight at the bottom, and dawn to the right (the Earth rotates anticlockwise). The auroras lie in a roughly circular region surrounding the pole, the weak emission seen at the top of the plot being due to ultraviolet light from the Sun which has been scattered by the Earth's upper atmosphere. SuperDARN radar flow 'arrows' and streamlines are superposed in black, showing a somewhat distorted twin-cell Dungey-cycle flow in the polar ionosphere (altitudes of a few hundred km), while the magnetic disturbance observed on the ground is shown by the red 'arrows'. The latter have been 'rotated to show directions corresponding to the ionospheric flow, opposite to direction of the ionospheric current, assuming that the currents are wholly Hall currents. The diagram at bottom right shows the direction of the upstream interplanetary magnetic field, where the red arrow shows the field direction as seen by an observer looking at the Earth from the Sun, as measured by a spacecraft in the solar wind upstream of the Earth's magnetosphere.
From Grocott et al. , Ann. Geophysicae, 20, 1577, 2002

Returning to Earth, the Dungey-cycle flow, when present, is communicated from the magnetosphere to the ionosphere (at altitudes down to ~100 km) by Alfvén waves travelling along the field lines. This produces a twin-vortex flow in the ionospheric plasma in which open field lines move over the polar region from the dayside to the nightside (see Fig.11), while closed field lines return sunward at lower latitudes via dusk and dawn. The overall cycle time is typically six to twelve hours. Over the last ten years these motions have been continuously monitored by a growing network of purpose-built radars (the SuperDARN network) which probe the northern and southern polar ionospheres at altitudes of a few hundred kilometres, and can produce a map of the flows in these regions every minute or two. One such flow map is shown in Fig.14, corresponding to an interval during which a polar substorm was in progress. The black 'arrows' indicate the flow directions measured by the radars, while the black lines show the overall streamlines of the flow computed from them. Such flows are, however, resisted by the friction produced by collisions between the ions in the plasma and the neutral gas atoms in the upper atmosphere with which they mingle. This causes a deflection and slowing of the ion motion in the lower ionosphere where the collisions are most intense, while the motion of the lighter electrons remains almost unaffected. The consequent differential drift of ions and electrons constitutes an electric current which is generated directly by the flow.

The polar magnetic disturbances measured by Birkeland and his successors are in fact the magnetic effects produced by the varying electric currents generated by Dungey cycle flow in the ionosphere. Part of this current, the Hall current, flows around the plasma streamlines such as those shown in Fig.14, but in the direction opposite to the flow since it is carried by the motion of negative electrons. These currents are responsible for much of the magnetic effect seen on the ground. Birkeland's 'equivalent current' arrows shown in Fig.6, for example, indicate the presence of an eastward ionospheric plasma flow (and hence a westward current) across Iceland, northern Norway, and Novaya Zemlya during a magnetospheric substorm. The same eastward plasma flow is indicated by the large red arrows seen near midnight (00 MLT) in Fig.14, which were similarly generated from ground magnetic disturbance data, now rotated into the direction of the flow assuming that the magnetic effect is due wholly to Hall currents. The time and effort required to generate one such map in Birkeland's day was immense. Today, magnetic disturbance and ionospheric flow data are generated routinely worldwide every minute or so, and are combined together to produce maps such as that shown in Fig.14 in near real-time via web-based links.

Another component of the ionospheric current is called the Pedersen current, which characteristically does not close within the ionosphere. Instead, after flowing across the ionosphere the current flows up and down the polar magnetic field lines above the ionosphere, and closes in the magnetosphere and solar wind beyond. In effect, the magnetosphere and solar wind play the role of the 'generator' in this current circuit while the frictional ionosphere is the 'load', and the electric current flows between them along the magnetic field lines. The ionospheric Pedersen current and the field-aligned currents combine to produce a generally small magnetic effect on the ground, but a large magnetic effect above the ionosphere, where the field lines become 'tilted' as though being dragged by the solar wind and magnetosphere through the frictional ionosphere. Evidence for the existence of this current circuit was found in magnetic data from spacecraft flying above the polar ionosphere in 1966, though it was not recognised as such at first[16]. Generally, the field-aligned currents are carried by light electrons rather than heavier ions, so that downward currents, for example, are carried by electrons flowing up out of the dense cool ionosphere. Upward currents, however, are produced instead by hot electrons flowing down the field lines from the magnetosphere to the ionosphere, and very often these have to be accelerated along the field lines by electric fields above the ionosphere in order to make enough current flow. When these accelerated electrons hit the top of the atmosphere they excite the atoms and make it glow, forming a 'discrete' or curtain-like aurora, such as those photographed by Störmer in Fig.4.

Birkeland was therefore correct in his supposition that 'discrete' auroras are intimately linked with current flow along the magnetic field lines, though the reasons for the existence of the current, and its connection with ground magnetic disturbance, lie in a realm of physical complexity far beyond that which could be contemplated in his day. Rather, the auroral situation is more akin to the field-aligned current 'discharges' envisaged by Alfvén. Alfvén was also correct in his suggestion that the main phase of magnetic storms is due to the current produced by the drift of hot plasma injected into the inhomogeneous magnetic field of the Earth. However, with his rejection of the notion of the Chapman-Ferraro magnetopause, the direction in which he envisaged the plasma to be injected, directly from the Sun (as in Fig.10), is the opposite of actuality. As shown in Fig.11, the hot plasma is in fact injected from the tail towards the Sun as part of the Dungey cycle. The existence of the magnetopause was, according to Alfvén, 'the biggest surprise of my scientific career' (as he once told the author during a visit to Dungey's group at Imperial College, London, in the early 1970s). Alfvén also rejected the significance of magnetic reconnection and Dungey's ideas about the solar-terrestrial environment, calling them 'PSEUDO THEORIES' (in capital letters!) in print[15]. Given the predictive and unifying success of Dungey's concepts it seems doubtful that many would agree with him today.

We can now watch the evolution of auroras over the whole polar region using cameras on high-flying spacecraft, and study how they relate to the transport of magnetic flux in the system in the Dungey cycle. Typically it is found that when the interplanetary magnetic field turns southward, open field lines are created at the magnetopause and transferred to the nightside tail over intervals of around an hour, resulting in an expansion of the region of open flux in the ionosphere, and an equatorward motion of the auroras. Then the tail current system becomes unstable (exactly how and why is still a matter of research), resulting in the rapid closure of open flux in the tail and its collapse back towards the Earth, as mentioned above. This collapse generates large electric currents flowing between the tail and the ionosphere which produce large magnetic effects on the ground on the nightside (Birkeland's 'polar elementary storm'), as well as bright discrete auroras which move rapidly poleward. The collapse also heats and compresses the magnetospheric plasma, producing the hot 'plasma sheet' population shown by the red dots in Fig.11. This plasma too flows down the field lines into the ionosphere, producing bright pulsating 'diffuse' auroras equatorward of the discrete forms. A snapshot of the aurora during such a disturbance taken in ultraviolet light by the VIS camera on the Polar spacecraft is shown in Fig.14, on which the SuperDARN radar data discussed above has been superposed. The aurora forms a ring around the pole (mapping basically to the 'red dot' region of Fig.11), with the brightest emissions occurring near midnight at the bottom of the image, mapping to the field 'collapse' region in the tail. The auroral behaviour during these events corresponds to Störmer's 'succession of events in a common aurora borealis display', mentioned above, which the Birkeland-Störmer theory could not directly explain.

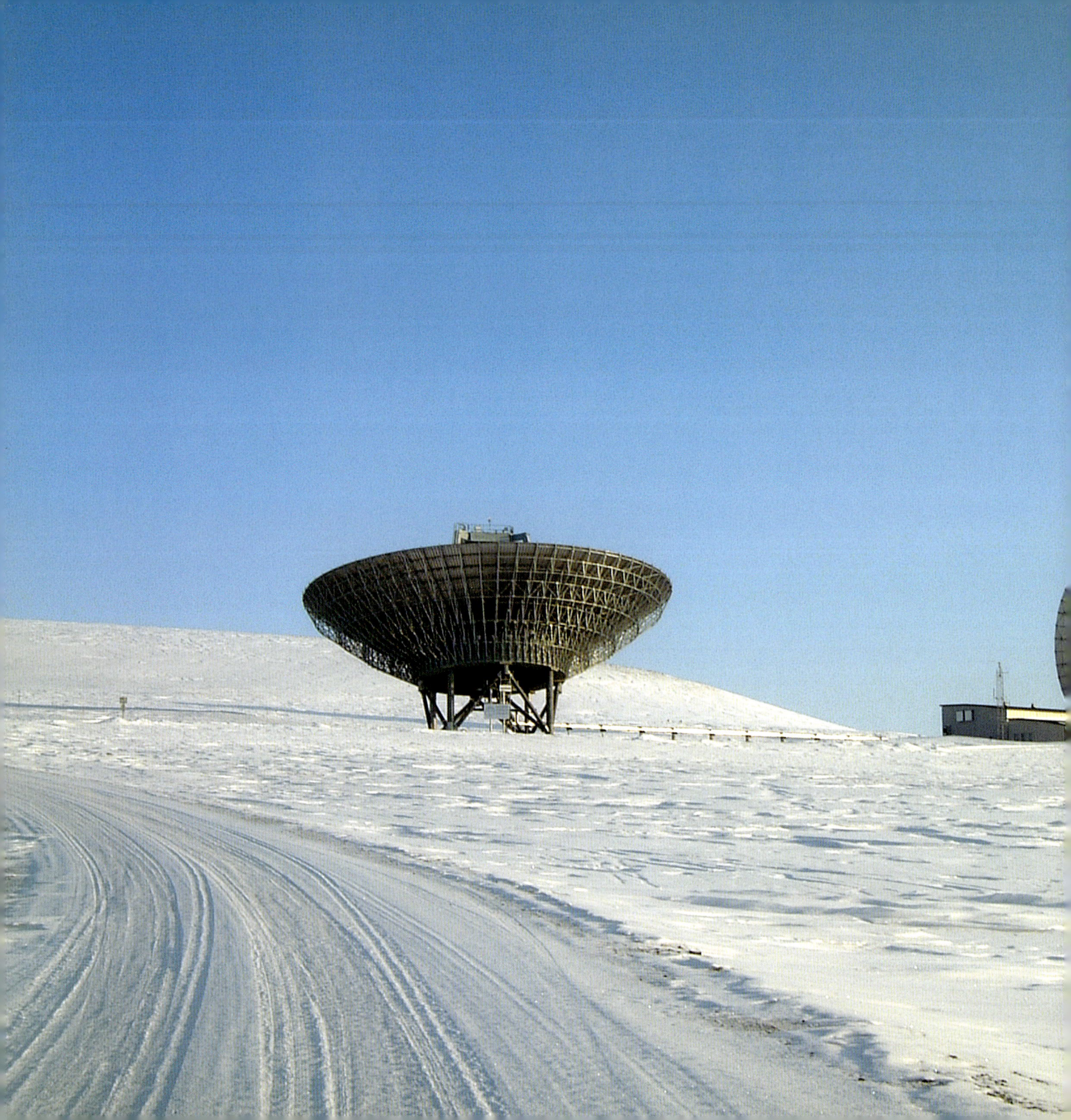

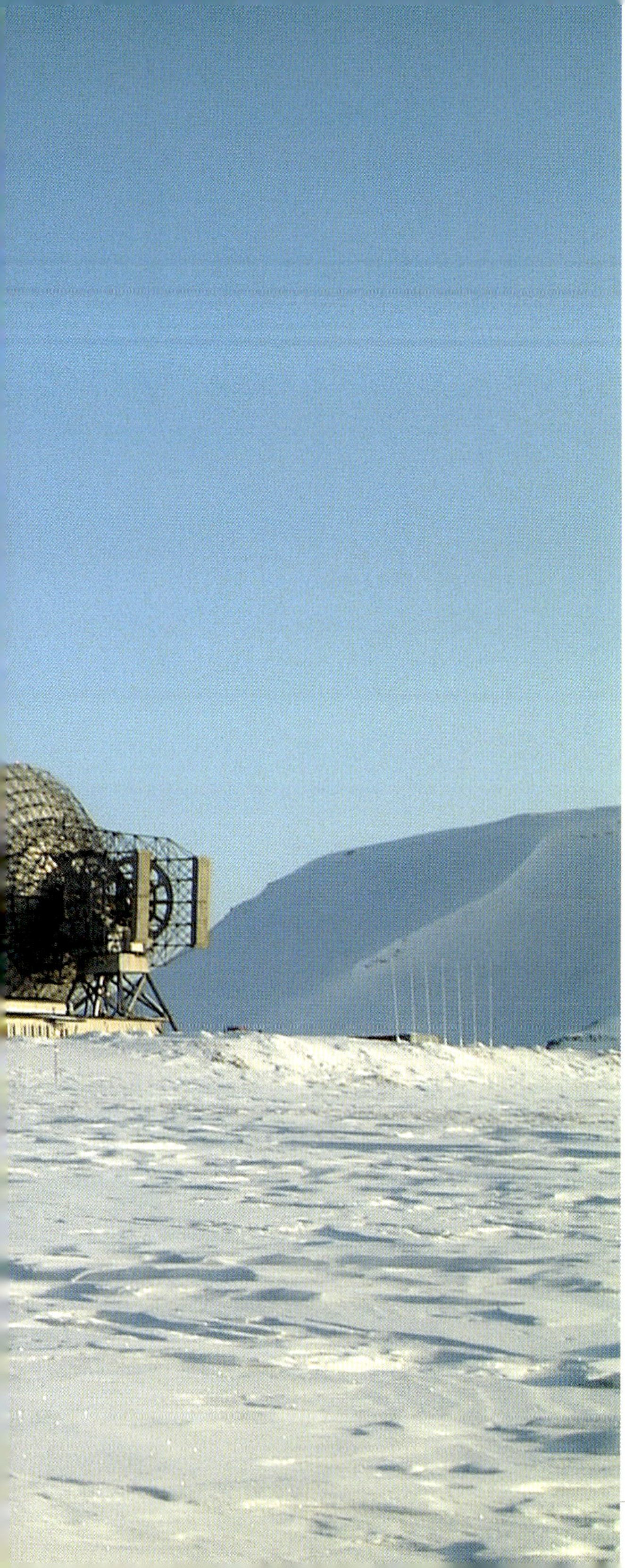

Fig.15
Photograph of the EISCAT incoherent-scatter radar on Svalbard, consisting of a 32m diameter fully-steerable dish (right), and a 42m fixed dish (left). This ionospheric radar system operates at a frequency of 500 MHz, so that to a first approximation the radio signals transmitted from the dishes pass straight through the whole ionosphere. However, the electrons in the ionosphere produce a weak backscattered return signal, detected using the same dishes, from which the properties of the plasma, the density, temperature, and flow speed, can be derived along the radar beam. The EISCAT Association also operates similar systems in northern Scandinavia.
Photograph courtesy of the EISCAT Association

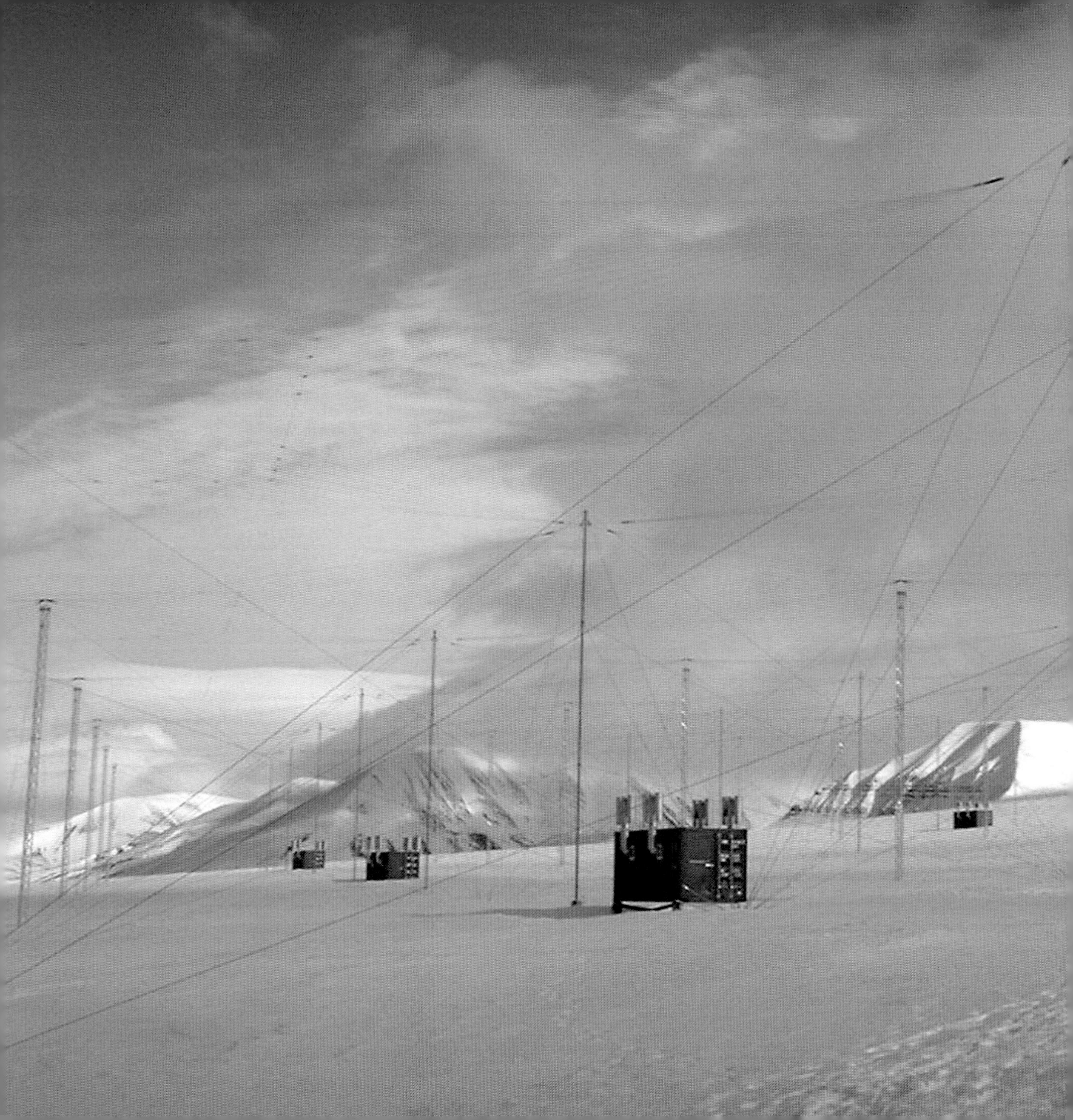

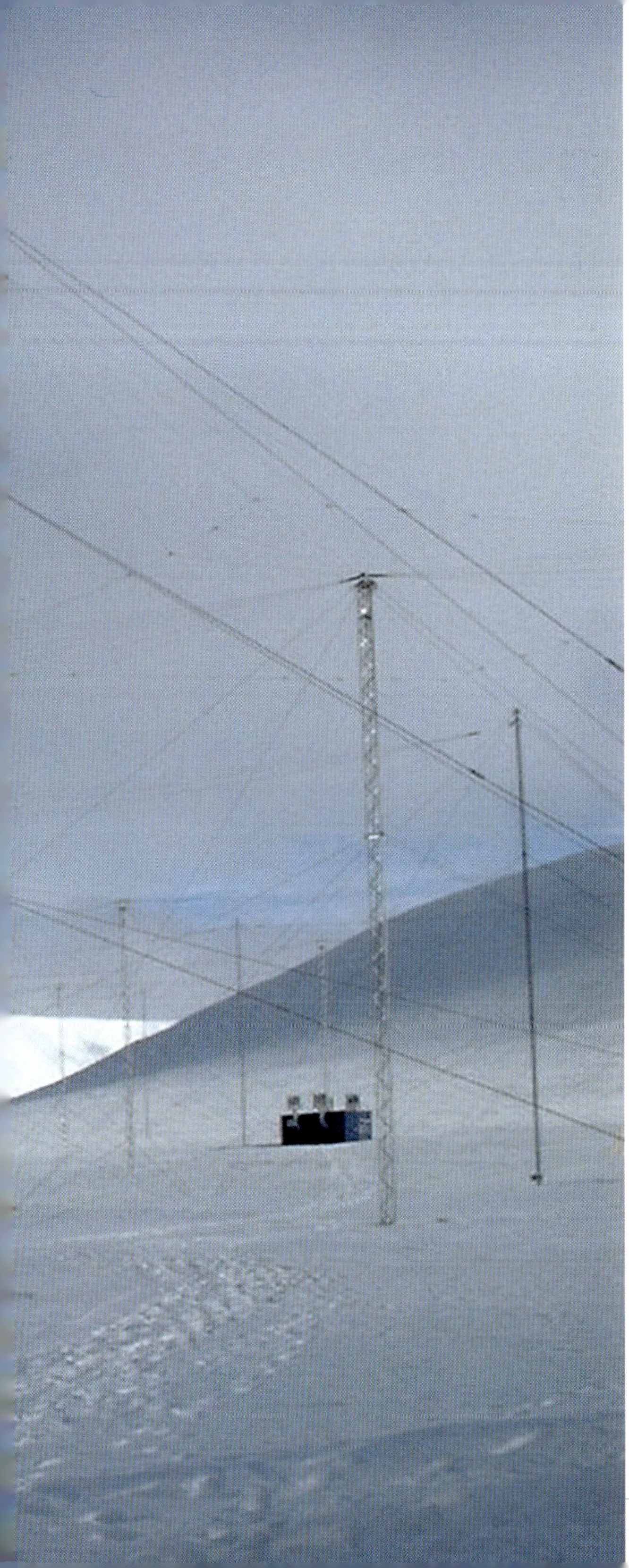

Fig.16
The new SPEAR high-power ionospheric radar system was built by the Radio & Space Plasma Physics Group at the University of Leicester with support from the PPARC research council, and was deployed on Svalbard in summer 2003 on a site adjacent to the EISCAT Svalbard Radar shown in Fig.15. The image shows five of the six shipping containers, complete with snow-proof cooling 'chimneys', which house the radar transmitters. These feed radio power to the antenna wires supported by a series of masts, which transmit high-power radio waves vertically upward into the ionosphere. Part of this radio energy is absorbed, creating waves in the ionosphere that can be measured by other radars, and creating temporary changes in the ionospheric temperature and density that can be used as probes to investigate ionospheric plasma processes. Photograph courtesy of Richard Poyner, RSPP Group, University of Leicester

As we have seen, the unravelling of the complexities of the Earth's plasma environment over the past forty years has involved the development of major multi-national scientific infrastructure, both in space (Figs.12 and 13) and on the ground (Figs.15 and 16), the data from which have provided the necessary stimulation and testing-ground for subsequent development of theoretical ideas.

We now know that most of the material in the universe is in the plasma state, either because it is very hot like the Sun's atmosphere (the corona), or because it is irradiated by ultra violet light or X-rays, like the Earth's upper atmosphere (the ionosphere). As Birkeland first foresaw 100 years ago, and Alfvén and others after him, electromagnetic phenomena play a central role within the cosmos. However, understanding the complex structures of such systems and discovering how they behave has been a difficult task, since plasmas are good electrical conductors that interact strongly with the electromagnetic field, while we happen to live in one of the few places in the universe which is sufficiently cool and insulated from irradiation that it is a reasonably good electric insulator. As the astronomer Arthur Eddington (1882-1944) remarked: 'We are bits of stellar matter that got cold by accident, bits of a star gone wrong'.

Our terrestrial thinking about electromagnetism, based on the experimental and theoretical discoveries of the nineteenth century, is strongly coloured by this fact. In order to understand the electromagnetic workings of the cosmos, starting with the solar system, we have had to learn differing ways of thinking that apply to plasmas, though still founded on Maxwell's equations and Newton's laws. This process started with Birkeland, though it was far from the possibility of completion at that time given the conceptual and technical means at his disposal. Much has been written[16] about the subsequent history and the interactions between the principal players, the scientific hegemony enjoyed by Chapman during the middle years of the twentieth century, its negative effect on Alfvén's career (crowned nevertheless by the Noble Prize for Physics in 1970), and the neglect of Birkeland and Störmer's ideas for fifty years (which a reading of contemporary works shows to be untrue – it was never forgotten). In the end it turns out that each of these pioneers contributed a significant part to modern understanding, now unified within the framework proposed by Dungey. Today we value the positive contributions that each of them made, even if at the time they found it difficult to value each other.

Acknowledgements.

The author would like to thank all the members of the Radio & Space Plasma Physics group at Leicester who participated in various ways in the *Little Earth* project, specifically Enrico Arnone, Lisa Baddeley, Emma Bunce, Stuart Crooks, Atousa Goudarzi, Adrian Grocott, Paul Hamadyk, Steve Milan, Jon Nichols, Richard Poyner, Hannah Scoffield, Gabby Provan, Jim Wild, Darren Wright, and Tim Yeoman. Professor Alv Egeland is also thanked for his help with Birkeland-related images. SuperDARN- and SPEAR-related research at Leicester is supported by the UK Particle Physics & Astronomy Research Council.

Bibliography

1 W. Gilbert, De magnete, magneticisque corporibus, et de magno magnete tellure: Physiologia Nova: London, 1600.

2 J. C. Maxwell, A Treatise on Electricity and Magnetism, Clarendon Press, Oxford, 1873.

3 An excellent summary of Birkeland's life focussing mainly on his scientific achievements may be found in A. Egeland's article, Kristian Birkeland: The man and the scientist, in Magnetospheric Currents, ed. T. A. Potemra, Geophysical Monograph 28, p.1, American Geophysical Union, Washington, 1984. For a broader account of Birkeland's life the reader is referred to Lucy Jago's biography The Northern Lights, Hamish Hamilton, London, 2001.

4 A full account of Birkeland's experimental researches can be found in the report by T. Brundtland, The laboratory work of Professor Kristian Birkeland, University of Tromso, 1997.

5 For a discussion of early Scandinavian accounts of the auroras, as well as later developments including Birkeland's work, see A. Brekke and A. Egeland, The Northern Lights: Their Heritage and Science, Grondahl Dreyer Publ., Oslo, 1994.

6 K. Birkeland, Sur les rayons cathodiques sous l'action de forces magnetiques intenses, Archives des Sciences Physiques et Naturelles, Vol.4, p.497, Geneva, 1896. [Only the primary French-language publication is given here. For the other references see Ref.4 above.]

7 C. Störmer, The Polar Aurora, Oxford University Press, London, 1955. See Part I, Ch. III-VIII for a full discussion of the subject of auroral height determination using photographic methods. See Ch. IV p.40 for a discussion of early attempts to measure auroral heights. See Part II, Ch. I-VIII for a discussion of work on trajectories in a dipole magnetic field. See Part II, Ch. VII for a discussion of post-Birkeland terrella experiments. See Part II Ch. XI Sect. 61 for a description written by Alfvén of his theory of magnetic storms.

8 K. Birkeland, Expedition Norvegienne de 1899-1900 pour l'étude des aurores boreales, Resultats des recherches magnetiques, Videnskabsselskbets Skrifter, I Math Naturv., Christiana, 1901.

9 K. Birkeland, The Norwegian Aurora Polaris Expedition 1902-1903, Vol. I, First Section, pp.1-315, H. Aschehoug and Co., Christiana, 1908.
K. Birkeland, The Norwegian Aurora Polaris Expedition 1902-1903, Vol. I, Second Section, pp.319-801, H. Aschehoug and Co., Christiana, 1913.

10 S. Chapman and J. Bartels, Geomagnetism, Vols. I & II, Oxford University Press, London, 1940. For discussion of the Chapman-Ferraro theory of magnetic storms see Vol II pp.853-869. For comments on the relationship with Birkeland-Störmer theory see e.g. Vol II pp.834 and 885.

11 S. Chapman and V. C. A. Ferraro, A new theory of magnetic storms, Part I, The initial phase, Terr. Mag. Atmosph. Electr., 36, 77, 1931.

12 H. Alfvén, A theory of magnetic storms and of the aurorae, I, Kgl. Sv. Vet. Akad. Handl., ser.3, 18, no.3, 1939.

13 J. W. Dungey, Cosmic Electrodynamics, University Press, Cambridge, 1958.

14 J. W. Dungey, Interplanetary field and the auroral zones, Phys. Rev. Lett., 6, 44, 1961.

15 H. Alfvén and A. Egeland, Auroral Research in Scandinavia (First Kristian Birkeland Lecture), Norwegian Academy of Science and Letters, 1987. [The author had the opportunity of adding his own comments to this story when he had the honour to present the 8th such Lecture in 1995.]

16 For the story of the discovery of field-aligned currents flowing between the magnetosphere and ionosphere see A. J. Dessler, The evolution of arguments regarding the existence of field-aligned currents, in Magnetospheric Currents, ed. T. A. Potemra, Geophysical Monograph 28, p.22, American Geophysical Union, Washington, 1984. This article also discusses the work of Birkeland, and the relationships between Chapman, Alfvén, and Störmer, according to the viewpoint of its author.

Little Earth Publication Contributors

Professor **Stanley WH Cowley** is Head of the Radio and Space Plasma Physics Group in the Department of Physics & Astronomy at the University of Leicester. The RSPP Group is at the forefront of research into the interaction of planetary environments with the solar wind. The principal science goal of the Group is the investigation of the outer plasma environments of Earth and the planets, and how they are shaped by the interaction with the Sun and the outflowing solar atmosphere. **James Flint** is an author whose novel, *52 Ways To Magic America* (2002: Fourth Estate, UK) won the Amazon.co.uk Bursary Award (2000). He wrote *The Book Of Ash* (2004: Viking), based on the true story of an artist trying to make art out of nuclear waste. One of his stories, *The Nuclear Train*, was adapted for television by the filmmaker Dan Saul, and recently shown on Channel 4. **London Fieldworks** was co-founded in 2000 by artists **Bruce Gilchrist** and **Jo Joelson** as an umbrella organisation for their cross-disciplinary collaborations. **Gustav Metzger** was a co-founder with Bertrand Russell of the Committee of 100, the anti-war protest group. He convened the now legendary *Destruction in Art Symposium* in 1969, and proposed the first Art Strike in 1974. *Metzger*, a documentary by Ken McMullen was commissioned by Arts Council England as part of the Pioneers in Art and Science series. Dr **A L MacKinnon** is a member of the Astronomy and Astrophysics Group in the Department of Physics & Astronomy, University of Glasgow. He is Access Co-ordinator for the Department of Adult and Continuing Education (DACE) and is currently a Board Member of the Community of European Solar Radio Astronomers (CESRA). **Dugal McKinnon** is a composer who teaches instrumental and electronic composition, analysis and 20th-century music history at the Victoria University of Wellington. Following the award in 2003 of a Victoria University Research Fund grant for Creative Interface, he is undertaking a practice-based research project that explores compositional technology in interdisciplinary collaborations. **Marjory Roy** joined the UK Met Office in 1961 and in 1981 was appointed as Superintendent of the Edinburgh Meteorological Office, the climatological office for Scotland, which holds the Scottish weather archives, including the Ben Nevis observations. She retired in 1990, completed an M Phil in meteorology at Edinburgh University in 1995 and became Honorary Secretary of the Scottish Center of the Royal Meteorological Society in 1997. **Jeni Walwin** is a writer, curator and Project Director of Art at the Center, Reading – working with artists to influence city center developments. She is a selector for *ArtFutures*, an annual selling show for the Contemporary Art Society. **Tracey Warr** is a curator and writer specialising in body art and site specific art. She edited *The Artist's Body* (Phaidon, London, 2000) and is the Director of Arts & Cultural Management at Dartington College of Arts.

Little Earth Acknowledgments

Allenheads Contemporary Arts, Ben Nevis Distillery, Marcus Allard, Chris Allen, Davie Austin, Elisabeth Beard, Alan Boldon, Patricia Bland, Asgeir Brekke, Broslum, Paula Brown, Alistair Brotchie, Laura Brun, Terje Brundtland, Andrew Conway, Paul Doherty, Ramsey Donaldson, Joyce Donaldson, The Exploratorium, Bronac Ferran, James Flint, Malcolm Fraser, Tom Freshwater, Bertie Fritsch, Alex Gillespie, Mary Gillespie, Mike Guest, Peter Guttridge, Sian Hamlett, Truls Lynne Hansen, Jonathan Hart, Headlands Center for The Arts, Nigel Hawkins, Ed Holloway, John Hutchison, Thomas Immel, Laura Immel, Jon Iredale, Brett Isham, Lucy Jago, Anja Jenssen, Rob Kelly, Yarrow Laser, Laurie Laser, Darryl Laser, Norbert Lieder, Lochaber Junior Gaelic Choir, Lochaber Pipe Band, Lochaber Reel and Strathspey Society, Helen Lucas, Chris Lucy, John Newcombe, Fiona Marwick, Alec MacKinnon, Ali Macdonald, John MacLennan, Babs McCool, Mark Miodownik, Cllr Olwyn Macdonald, Dugal McKinnon, Gustav Metzger, Barry Moss, John O'Kane, Judy O'Rourke, Marit Hætta Øverli, Judith Palmer, Chris Pearman, Tanya Peixoto, Jan Pettersen, Arvid Petterson, Miranda Pope, Helen Ratcliffe, Clare Reddington, Iain Robertson, Mary Robinson and the Estate of CTR Wilson, Marjory Roy, Michael Sanders, Hans Christian Søborg, Science Museum London, Scottish Mountaineering Club, Alan Smith, Katie Smith, Ian Thomson, Jane Thomson, Mick Tighe, Tony Van Eyken, Jeni Walwin, Tracey Warr, Jane Wess, West Highland Museum, Tony White, David Wilson, Pam Winfrey, Jules Wright, Professor Stanley Cowley and members of the Radio and Space Plasma Physics Group at the University of Leicester: Enrico Arnone, Stuart Crooks, Atousa Goudarzi, Emma Bunce, Lisa Baddeley, Adrian Grocott, Paul Hamadyk, Steve Milan, Jon Nichols, Richard Poyner, Gabrielle Provan, Professor Terry Robinson, Hannah Scoffield, Chris Thomas, James Wild, Darren Wright, Tim Yeoman.

The Little Earth project gratefully acknowledges funding from:

Arts Council England, Scottish Arts Council, Lochaber Enterprise,
The Highland Council, University of Leicester, Arts & Humanities Research Board and Arts Council England through the Arts and Science Research Fellowships Scheme.
Support in kind from Alta Museum, West Highland Museum, Wapping Project, Nevis Partnership and Alcan.

Scottish Arts Council LOTTERY FUNDED · ALCAN · A·H·R·B arts and humanities research board · University of Leicester · Lochaber Enterprise Iomairt Lochabair · The Highland Council Comhairle na Gàidhealtachd · Arts Council England LOTTERY FUNDED

Published by London Fieldworks Limited, in an edition of 500.

Editors: Bruce Gilchrist and Jo Joelson
Design: Chris Lucy
Printed in the UK.

ISBN 0 9549497 0 6

British Library cataloguing-in-publication data. A catalogue record of this book
is available from The British Library.

London Fieldworks Limited
Unit 11, Ellingfort Road
London E8 3PA, UK
t: +44 (0)20 8985 6340
f: +44 (0)20 8985 6340
e: info@londonfieldworks.com
http://www.londonfieldworks.com